Jas   Fiza   Science

(Complete)

[Execution of Time (Novel), 2nd Moon (Short Stories), Nature Summons (Poetry) Three in One.]

By: Jas Fiza

Three Books in one

Execution of Time (Novel)

2nd Moon (Short Stories)

Nature Summons (Poetry)

As

Jas Fiza Science (Complete)

By: Jas Fiza

Publishing Year: 2013

Price: & 18.00

Introduction of writer

Name: Jaspal Singh Randhawa

Literary name: Jas Fiza

Birthplace: Village Fateh Jalal

Teh & Distt. Jalandhar,India

Present Address: 5211 Fawn Hollow Way

Antelope, CA, USA

Phone: 916-745-4946

Cell: 916-596-5170

Email: jasfiza@msn.com

Other books of the author:

Execution of Time	Novel (English)
2nd Moon	Short Stories (English)
Mita Dita Hava Ne Nan	Gazals (Punjabi)
Kalave De Hava	Laghu kathawan (Punjab)
Lash Bol Rahi se	Laghu Kathawan (Punjabi)
Raat ke Aakhegi	Gazals (Punjabi)
Bharti Gawahi Kanoon	Khoj Patar (Punjabi)
Che Sodh De Lorh	
Varmian Che Vasde lok	kavtavan (Punjabi)
Bharat De Jakhmi Chirhia	Parhchol (Punjabi)
Sabdan Da Chola	Kahanian (Punjabi)
Javabdehi	Lambi kavita (Punjabi)
Javab Talb	Lambi kavita (Punjabi)
Nature Summons	Poetry (English)

1ST Book

Execution of Time

By

Jas FiZa

(Novel)

Execution of Time

Novel

By

Jas Fiza

Publishing Year: 2010

-1-

     The moderator heard that some souls were going to discuss life. At first, he laughed at them loudly. Then he said, "I have no concern with their discussion, but they have to face the realities that they are not aware of and those realities are above the reality of life. Man is racing towards destruction. Man has interfered in the routine work of nature and earth. Man has gained the power to extract the strength of the earth at large but he has no power to return it, in the way that is necessary to maintain the balance of its regular functions. Accordingly, the man has taken out the strength of the earth in large quantity and harmed the ability of the earth to work as normal. Now he is unable to restore that strength to earth. Man has weakened the earth and upset its balance.

     This is not happening for the first time over the earth. This has happened on the earth repeatedly. Again and again it has happened like this. Man has not gained this power for the first time. It was written in the tomes and sacred books many years ago, before there was such progress, that there would be big bombs and missile systems over the earth in the present time. The names of those bombs were different from the present ones. The names, like agan ban, agan shastar were the same as those of the missiles being controlled by man over the earth, when meeting their targets in the air. Drones were also used freely at that time in the wars. Man freely used the air space at that time. Satellites were also fully developed. It is written in the sacred books that a blind king sitting at home was hearing from his officials in the field every moment of the war, watching on television via satellite. Satellite, missiles, bombs developed many years ago before the present progress—where are those systems? Where is that progress? Destruction after progress, destruction after progress, how many times this will happen? Earth will again discipline man to keep him without knowledge. When? The answer is, very soon. A few years ago, when present progress did not exist, it looked as if those miracles would not ever exist. Now the reality has become perception. Earth will take action very soon. Mother Nature will keep the balance of earth in the planets.

Destruction is at the door of man's knowledge. Less than a century ago there was nothing.

Man raced towards destruction very fast and will knock on the door of destruction very soon.

Man again has defeated god. Man has discovered bombs, which can finish the whole world in seconds. God has again become helpless to stop man from doing that. Who will take action first, god or man? Who knows? Man is adamant about challenging god. Who will win this war?

When the earth takes a deep breath to recover its strength, to maintain its balance, what will happen?

Before the moderator said anything more, one unknown soul told him that the subject of the souls was not only man's life. Their discussion concerned the whole of creation.

The moderator said, "It is of no matter, what the subject is. No one on the earth except man is responsible for the cause of destruction. When the buffalo dies, so too, does the louse."

_2_

"As you know, we, the souls, are in heaven; hence, we should not pursue desires. Desires, wishes and greed—all are the needs of living things. As we have already left the somas, it's useless to bother about life, from which we are now liberated, and which the red soul wants to discuss," the multicolor soul said.

"It's not our desire, wish or greed; we only want to see the power of time, which prevails in the world, and now we can see and observe just that. Nothing is in the hands of the human being, which is dealing with all creatures, nature and life, as humans are the only controlling power on the earth. There is nothing in front of the human being to show that its acts and deeds are fare and that it operates without partiality. Otherwise, with the acts and deeds of human beings affected adversely all the nature, creations, and all the life. So I think this is not a throwaway idea." The blue soul supported the red soul's view.

"In a real sense, we should not discuss those things that are not in our power to change however, this issue is different, and is not constrained by this view. We are just discussing life, some past moments of our lives, and we are well aware of our limits. So there is nothing wrong with discussing the view that comes before us," the green soul said.

"We know all aspects and realities of life, as we have the experiences of our past, which we have gained while we were alive. So, why should we want to discuss that thing that we have lost and have no concern with in the future?" The yellow soul expressed its view.

"We have concern, even if we are souls and cannot see beyond our limits. But ultimately, our place to live is life in the future. We left life recently in the past. There is no harm in discussing the thing that was our past and that we can achieve in the future. Moreover, we're only putting forward our views from our own perspectives, so there is nothing wrong in discussing them." The red soul put her view forward emphatically before the others.

"Why we are going to discuss this?" the yellow soul asked.

"We should not make this an issue. This should not be the issue before us, to discuss why, how, what, when, where or who. We are souls and not living things now. We should forget all our personal agonies. However, I don't think we are doing any wrong in discussing life." The blue soul suggested.

"Life comes and goes. How many lives have been on the earth and what they did in the past, how many are living on the earth at present and what they are doing and how many will come and what role they will play...even if we had that knowledge, it would still be in our past. Hope for the future is useless for souls. I don't think we should touch those issues that are useless for us." The yellow soul again tried to convince the other souls.

"I was a man when I left my body. I think you have to agree that I have more experience than you of life, which may be good or worse for us, but I am of the view that we should know more about the life we experienced recently. I know the experience of the hard life of the yellow soul, who died as an ant, but I think she is unnecessarily reluctant to discuss life." The red soul said.

"I left my body as a parrot, the blue soul as a dog and the multicolor soul as a butterfly, but I think we are not very much tied to our recent past experience of life. We wish to discuss life as life, not as personal experience. We should leave aside and forget all the contradictory views we have stated earlier, as we are going to start the discussion about life," the green soul said, as it had been given the authority to decide on the primary discussion.

"We should appreciate each other's views and not linger on the matter, keeping in focus the importance of the task," the multicolor soul said, giving its full consent for discussion.

The yellow soul said no more and looked as if it also agreed with them. All the other souls were silent. It appeared clear that they all were ready to discuss life.

"Life is a song, which everyone writes with his blood and has to sing on his soma during life. if you fail, the poetry will die." The red soul started the discussion.

"Life is a secret of happiness but pain exposes it. So why you hide the pain and happiness and make them strangers in their own home?" the green soul wondered aloud.

"Life has meaning, but who tells us the meaning of life? Some have made it a mystery and started to worship it and forget the meaning of life." The multicolor soul entered the discussion.

"Life is a war in which death is certain. But why should we die a cowardly death? Why don't we fight the war to conquer death?" the yellow soul asked, now taking part in the discussion.

"Life is a valuable thing. We should increase the value of life by sacrificing it, rather than dying from being trampled under atrocity," the red soul said.

"Life is very precious and prestigious. Don't make it cheap by your behavior—always behave like a brave man," the blue soul said.

"Don't depend on others to decide your life for you. Your first right is to decide your life. Take decisions by yourself and have self-respect," the yellow soul said.

"Don't worship for your satisfaction the people who, at the time of deciding about life, accepted death. Adopt their path of reality to give respect to life. They accepted death out of respect for life." the red soul said.

"Don't expect more from those who have sacrificed their lives for you; they will not come again to help you. Now, you have to do something," the multicolor soul said.

"Look forward. Do what you want to do in life and be the partner of nature, and not of the material things made from the natural sources by others. Nature will save and material things will ruin you," the yellow soul said.

"Man's next war may be for nature, to save life, but who knows whether he can win or not. Has man become able to win the war against nature? " The red soul expressed her worries.

"Yeah, I agree. Man has to decide which is more important for life, worldly things or nature, which plays a role for all creatures. If he thinks nature itself is a material, then I think the selfishness of man itself will kill him." The multicolor soul endorsed the views of the red soul.

"Life is not a commercial transaction. Save yourself from trade. Life is play. Play it. It's not a one-day game. Be prepared to dedicate your whole life to it," the yellow soul said.

"Love the water, you will never be poor. Love the air, you will never be defeated. Love the sky, nobody can measure you. Love the sun, spread light and sunshine all around and nobody will have the courage to touch you. Love the earth, which bears the sufferings of man brought on by man, but still is of the view that man is ignorant," the blue soul said.

"Put your faith in nature. Life depends not upon material things made by man, but fully depends upon nature, and this universe is the pillar of nature." The green soul added.

"When you are living in the lap of nature, don't think yourself weak. When you feel yourself weak, understand that your power is being used by others. So many weak things together make strength again," the yellow soul said.

"Life is not conjecture, it is reality. Sufferings created are not confusion; they are the reality. Recognize reality and gain freedom," the red soul said.

"Don't think there is any difference between the waves of the ocean and waves in your mind. The waves cannot be measured or stopped. Break the iron chains of slavery as the ocean might" the yellow soul said.

"Grow like trees, serving others. Live like trees, servings others. Fall like trees, serving others, so others will remember your services of life. Be like trees, which serve even after death," the green soul advised.

"Nature is not made only for the human being to fulfill his needs by controlling and possessing and exerting his power over it. Every living thing in the cosmos has a right to enjoy the full fruit of it. It's the foolishness of the human being who thinks material things made by him are the root of life. This is his self-created problem. The importance of life is in the lap of nature," the blue soul said.

"Man is using his life as a machine. Man is giving the too much importance to the material life, which is the main cause of his sufferings, by giving up the natural life. A day will come when man will understand the depth of the universe and nature, when he will be disappointed in the material things he has made from the natural sources by destroying the natural sources," the yellow soul uttered.

"Nature has given a soma to life, which is a beautiful gift of itself. No one can beat nature in this holy cause. There is no substitute for nature in creating the body of life. It's only nature that can do this job. Nature not only gives the soma to life, but also takes the responsibility to nourish it after that. The body is nature's beautiful present to life," the green soul said.

"The soma given by nature to life is the process of nature alone. Life has no source beyond nature to create the same process. The basis of nature cannot be matched," the red soul said.

"Ignorant man cannot be compared with nature; he cannot even think about it. If he is foolish enough to imagine that by any little success of his scant knowledge he is extending as far as nature and its planets, then even if he reached some of the planets and captured and possessed some of the planets, as according to his wish he has captured and possessed the earth by his power and will, even then his selfishness cannot rival nature.

"Man is too small before nature and man and his achievements are nothing before nature. His condition is pitiable before nature. Nature's greatness is unbeatable in the face of man's progress," the multicolor soul said.

"Life is a bubble on the water in the play of time and nature. Someone is coming, someone is playing and someone is going in the play of life," the blue soul said.

"Man believes that nature is the impresario of life. Man is winning and defeating every moment from himself in the world around him. Society and the system built on the pillars of knowledge help him to remain in the competition, but selfishness always intervenes to isolate him. Life depends upon nature, and society and systems made by man are the burning issues before him to solve. He becomes theist and atheist while solving these questions, but fails to reach a conclusion. Maybe life is unable to solve his self-created problems arising from his selfishness." A moderator appeared and said these words.

All the souls were surprised to see the moderator. The red soul said, "We don't need any moderator."

"We have not invited anybody to express or to give any opinion. We will not accept any type of interference in our views," the green soul said.

"So many times when going on the path of life, I could not identify life. I felt shy. Under the unnecessary burden of life I could not see life. Every time I tried, truth danced before me like death, the only reality of life. Crestfallen, I never tried to accept the truth. I have no courage to say that truth will prevail. I will come again and again to intercede in your discussion. Maybe I am too debilitated to say the truth. The falsehood of life only wins with the truth of death." The moderator spoke and disappeared.

"The only way to deal with the presence of the moderator is to hear him, but ignore him," the multicolor soul said. All were agreed on this view.

_3_

It's the sharp light. They saw the light coming towards them, unable to understand what was happening. Light was all around them.

In the light, their faces of past birth came out rather the shadows of their faces of past birth.

In the light, it became apparent that their attachment to the last birth on the earth was pressing them to discuss their selfishness like this. They were still tied with their unfulfilled wishes and acts to the earth.

Everything was clear and nothing was hidden now. It seems the light was there only to observe their reality.

In the light the souls were unable to move and for a moment they forget their presence, existence and identity.

After some moments everything was becoming normal except reality which alone had changed. The effect of the light separated them for-ever and they were not together now.

In the next moment they found themselves far apart from each other.

The red soul was in the north.

The blue soul was in the south.

The green soul was in the west.

The yellow soul was in the east end.

The multicolor soul was there alone now. The multicolor soul was not scared by this action, at all. Actually it was expecting something like this to happen because of the abnormal activities of all the souls. In spite of what was happening the multicolor soul did not changed its mind about discussing the idea promised with other souls.

Afterwards there was no light and everything was normal again, except for the separation of souls. The multicolor soul, however, was alone without the presence of its companions and was watching the creatures, full of life. It saw an old man standing near a flower. He was telling to the flower his experience of life, which he gained from time to time by living long life.

"I have seen life like fishes writhing in the net recently caught by the fisherman who is putting them out to sell in the market." he said.

"I have seen life like a bird recently trapped by the fowler who will sell it to the bird fancier after teaching it some manners." the old man continued.

"I have seen life like waves left behind by the ships and boats on the water which will soon disappear into the water and all will as if nothing had happened there." he was saying.

"I have seen many storms of life."

"I have seen the light and dark of life." It looked as if the old man has disclosed himself from all corners.

The old man who had broken the flower from the flowering plant was silently looking towards it.

The plant saw the old man had no more to say any, so it scoffed, "Cowardly old man said nothing. What he said I can understand easily from the sweat stuck in the wrinkles in his face."

The multicolor soul was also listening with interest to their conversation interestingly but in the meantime it heard some cries in the air. These cries were not of souls but nobody was visible there for a moment. Soon it saw some lives hanging in between the life and death.

First it saw a tamed monkey who, because of his advanced age could not dance at the behest of his juggler owner, whom the juggler refused to keep with him anymore and had left to die, was crying, "Man has captured and taken into his custody the whole earth. I do not know how many kinds and races of birds, animals, trees and all other creations and creatures over the earth he has annihilated due to his own fear. More-over, man is claiming to be the developed form of the monkeys. But when I think about the behavior of man, I never wish to be a man. I do not want to be developed as a man anymore. What a cruel existence of life on earth as a man!"

The juggler immediately replied, "Man is also the victim of man who has taken everything into his custody and possession. His position Is no better than a bird in the cage, a monkey in the hands of a juggler, any other animal in the zoo or any other living being in the grip of man which is likely to be cut up for his food. If my old age is not secured, how I can provide security to any other? I am not mean but compelled by circumstances made by man and controlled by man. One has to go from the field and the game of life, unable to help others and living in peril. Monkey cannot blame the whole of mankind together."

The lamb tied to a post, his back legs tightly caught and pulled by a man to control his all movements, in his last struggle for life, looking towards the sharp edge of the sword in the hands of another strong man over his head, was crying, "I have no objection if I am being slaughtered for man's food. My mother never prayed for my long life as she knew my fate but I am worried about man who has made my meat so poor, staled from storage and destroyed its fine characteristics."

The lion was crying in the zoo, looking at the contaminated meat lying before him. On the other corner an elephant tied with iron chain was crying. Some birds were crying in the cages, which were unexpectedly defiled.

The multicolor soul found this boring. It had some sympathy with the crisis of life, but it was not her business. It looked away, towards the other side and other aspects of life.

White as milk, having a well mannered appearance, and a saintly face, moving slowly and calmly, keeping eyes half shut, nobody can doubt him. At first glance the multicolor soul also had the impression that he was a great devotee of god.

The multicolor soul also saw the water shiver, while white milky face was putting his orange feet in it. May be the water didn't trust the yellow beak and polite white face looking calm on the beautiful long legs and yellow webbed feet? May be nature understands calamity before it happens. May be some fishes innocently enjoying themselves will be finishing the last journey of their lives through the neat and clean beak up to the stomach of the heron in the next moment.

The multicolor soul saw big fishes eating small fishes, attacks of sea birds, animals and other sea living things from inside and outside, also the nets in the hands of some men, eyes of birds on the fishes and the philosophy behind the white faces.

The multicolor soul looked towards the bushes near the water where a tortoise was muttering. How many ideologies have developed in the past? The ideologies were not arrived at easily. Sometime ideologists or thinkers have to sacrifice their lives, bearing the atrocity of the powers, sometime they lives in the hell created around them by the atrocious people. The tortoise splayed his breast, not to show his muscles or the strength of his body but with a deep breath in grief. It is difficult to speak against power or a powerful system. Power never allows the truth to come out in the open air.

The tortoise sometimes slow and steady won the race from the hare. The tortoise many times saw the hare save his life from the hounds in a race of life and death, but today when he saw the hare in the cage of a man, he was muttering.

The multicolor soul then heard the conversation of two frogs. One was looking towards the sky.

The other said, "The sky is empty"

The first one said, "The sky is full of clouds, air, suns, moons, stars and so many other things of which we have no knowledge. The birds are also flying in the sky and I can understand the feelings of birds too."

The other one said, "We should not run after imaginations and feelings. Stars and moon needs night to twinkle and shine. Don't expect more from the fire of sun, which may burn you. Air is not your friend for-ever. One day it will leave you unattended and you will be no more. Earth is enough for us. Our life is on earth and after death we will be part of the earth."

The first frog said, "Nothing is dependency, take it a complimentary. Nights never disappoint you. The charming light of moon and stars when they embrace the earth at night and show their face to the creature, they enjoy those moments of love and affection and welcome the same. No doubt, the clouds are made from the water on earth but it is only to maintain the balance and save the lives on earth. Don't watch nature incompletely but take it creatively. We are living in the whole of nature, not only on the earth. If your feet are on the earth, then your body is in the sky. So don't think the sky is empty."

The frogs shared their views until they were put into the net of a merchant who will sell them for dissection by the scientist for experiments.

The multicolor soul now was on the mountains.

A monkey was saying loudly to the air, "Who polluted the atmosphere, endangering the life of every creature, and who is making the earth unsafe for all creatures? Are those monkeys, other animals or social animals? Who will explain then who is social?"

He was saying, "We want another chance to be social as our improved race has vehemently failed to keep nature unpolluted, and fit for living instead destroying the natural sources in the name of progress, which is harmful to all creatures. Moreover the progress of our improved race is impeding the efforts of the remaining monkeys to be social."

Another monkey started dancing over the views of his fellow and said' "If to be social is to accept boundaries then I request my friend to withdraw his plea to be social, as we cannot accept the boundaries."

A deer passing nearby said, "If you want to be social animals then first take sex education from me. All social animals have forgotten nature and use sex in the unnatural way. For the social animals sex is only a lust and nothing more. All social animals are living in a sea of lust. No other animal enjoys sex like the social animal. No animal use sex to fulfill their wishes of lust, rather maintains the law of nature."

The second monkey endorsed the deer and said, "The social animal has become the slave of his own laws and rules. He is the victim of his own laws. We have no laws and rules except the law of nature. We are unable to live in the chains of laws like social animals."

The deer said, "The social animal's bite is more dangerous than the snake's. The snake bites in fear only when there is danger to his life but the social animal bites to create fear to fulfill his malevolent and mean desire."

One more monkey came and said, "Man's knowledge was already limited to earth. Then he made more it limited by dividing earth into countries. Now he wants to expand his trade by winning other countries in wars. Still, he is taking up the burden of sociality on his shoulder with him. Animals never accept boundaries of their sociality like this."

The second monkey said, "Man is dying in search of truth. We live in truth and enjoy life."

The first monkey said, "I am not worried but it is a shame for man, as he knows names of some kings from his past and gains knowledge for his future from studying kingdoms. He does not know the pity and painful life of common people of the past or present and there is no change likely in the future. We do not remember our past and have no worry about the future. We don't make our life painful as man does."

The deer said, "Man is busy giving grief to man. Some persons think it the wish of god and some deny the existence of god but the curse is still on course towards the worst."

The second monkey said, "The tragedy of man is that whoever speaks against pollution, they kill him with pollution. Who educated people religiously explaining nature and its natural course to serve all and that nobody has the right to control it they first hanged him and then declared themselves his devotee. Whoever demands right of equality, they finished him with hate and discrimination. Whoever talks of poverty, they kill him with hunger. When one learns all the ways to kill, he becomes the successful ruler and metes out justice by means of democracy and laws which he maintains by force. His system becomes the benefactor of powerful people but the curse of the weak."

"Oh! What is this?" They said once.

"Fire in the jungle" Another horrible attack of man and they ran away hither and thither to save their lives.

The multicolor soul was now hearing the story of a pair of turbits was telling to a pair of pigeons.

"We were not turbits. We were pigeons like you. A man caught us and kept us together in the coop. Then my female pigeon laid eggs. At first the man released my female pigeon to fly in the sky. My female pigeon flew in the sky, but came back out of her love and affection for her eggs and me. Then infant birds hatched out of the eggs. During the fledgling period of our infant birds the man released us both to fly in the sky. We were bound to come back for our infant birds. Now everything was different. The values of life were changed for us. Life style was changed. Habits were changed. The meaning of freedom was changed. Caste changed, name changed and now we were turbits instead of pigeons." the male turbit said.

The pair of pigeon took a deep breath over the painful story of other's conversion.

"We were born as pigeons but will die as turbits. Our infant birds are born as turbits and will die as turbits. They don't know even the real meaning of freedom. They will never try to get freedom as they have become fully dependant on others. This is more painful for us. The man is now keeping us separate from our infant birds.

The man came towards that side and pair of pigeon flew away.

Now the multicolor soul was hearing the melancholy song of a cuckoo.

The multicolor soul already knew the pang of separation of the cuckoo. The cuckoo never made her own nest but always laid eggs in the nest of the crows in their absence. Crows can't judge anything from the different color and size of eggs. The female crow brooded over the eggs of the cuckoo together with her own eggs and hatched the young birds of the cuckoo, too. The cuckoo never brood and hatched her young.

The female crow came to know her foolishness and the cleverness of cuckoo when she heard the different voices of the young birds. She then threw out the young birds of the cuckoo from her nest.

It was the belief of the multicolor soul that the pang of separation was in the heart of cuckoo from birth.

A little ways away a man and a bird were twitting each other. The multicolor soul took interest in hearing them. They were not satisfied with each other's arguments. Both were trying to convince each other of their point of view. Both were explaining the logic of their plea with determination but could not reach a conclusion.

The man's plea was that man has discovered everything which is the peak of their progress to fulfill the needs of the human being. Man can even fly faster than the bird in their magnificent airplanes safely comfortably and with many facilities.

The bird disagreed with the man and trying to convince the man said, "The progress of which you are talking is the result of false wishes which man wants to fulfill by sacrificing all his happiness in life. Man has confounded himself badly with the worldly artificial things that he claims as achievements, but has lost his happiness of life in them. Man has become the slave of his wishes, which is the killer of his happiness. Man can't free himself from his false wishes. Man has entangled himself into worldly man-made things and is living the life of slavery to traders who discovered and controlled those things."

Man, being assured of his position, said more confidently, "We can enjoy different kinds of dishes as we have discovered. What have you? Even if you fell ill you have no doctor to cure. How you can compare your life with the life of man? Man is using his discovered things at every step and everybody is free to take the benefit of everything. "

The bird tried to conclude the matter and said to convince the man "Hey, what progress are you claiming? What is the benefit to you of all this? You can't compare your life with my life. You will never be able to get the air of progress. Your sweat goes to waste every time when it fells on the earth. Your blood is being sucked by the powerful in the name of trade.

The man laughed at him and said, "Don't try to confuse me about the problems of mankind. My discussion was only about your ways of living and my ways of living. Go away and save your life from the hawk or beware of the net fixed to catch you."

"I damn care of the future. It's the man who cares and dies worrying about the future, under the burden of the past and knows nothing about his present and how to save himself from this crisis." the bird said.

"Oh! Duffer! Don't you see the water is running into the fields of others through the broken enclosure? Are you aware that water is not irrigating our fields while you are enjoying the shade of the tree?" His owner caught him red handed absent from his duty.

He became nervous and looked upwards. The bird has already flown away.

He glanced at the tree with crooked eye.

The tree said, "Don't worry my friend, I shall remain here. Come again at any time to enjoy my shade by sleight to your owner, my shade is for you."

The multicolor soul saw towards another side. So many birds and animals were not worried about their livelihood or other troubles and problems given and created by man for them but they objected to the problems and troubles created by man for the whole world of creatures, and for the earth where they live and to water and air which are the main source of life for everyone.

Some were saying, "Man has developed currency parallel to the natural resources and runs his trade and business of natural things with money. He treats nature as his slave when he has only the right to use natural resources. Not only has man abolished their rights but the power of money creates suffering for the whole of mankind, too, by delivering it into the hands of a few men.

Now all the creations bear the sufferings of man's brutality and bullying behavior."

Some were saying, "Men have partitioned, captured, and taken into custody the earth and even the sky and water and shown possession over them. They make boundaries all over the earth and say it is their country. Man has captured all the natural resources under his control and fights wars with other powers to control even more natural resources of other countries for trade. Not only man is becoming his target and part of his bloodshed but all creatures have become his victims."

Some were saying, "Man is so mischievous that he compelled all creations and creatures to live under his control as slaves, but declared that he was providing freedom over the earth. Many birds, animals, men and other creatures bear the slavery of a few men's wishes, rule and trade. They are unable to save themselves, because of the power of ammunition and tactics of these few men."

Some were raising their voices, saying, "Man's progress and his mean wishes and callous desires have polluted the whole atmosphere of earth, air and water and his progress's ill will has made living hard on earth. His progress's lust has damaged the ozone-o-sphere even and earth is becoming warmer causing big problems for life on earth. Man's progress has given birth to so many dangerous diseases which are injurious to the whole of creatures."

The multicolor soul saw a child was explaining the definition of family. Another child whose family was ruined under the new philosophy was also telling something. Under one and the same crisis both were looking towards the stick of present and future standing on the bank of past which was ruining their life.

"A cool breeze was blowing. However it was the spring season. Trees swaying in the breeze looked like people extremely happy, enjoying the fair. Looking extremely happy, swaying without any specific reason. It looked as though the tender tops of the trees were saying something to each other by singing and dancing. Leaves on the branches of big tall trees were swaying as they were happy and enjoying the lovely atmosphere with other smaller trees. Some old leaves and flowers on the trees were watching seriously everything of new one and enjoying the moments. I also away from the struggle of life entered into the lovely moments with the trees. What a lovely moments they were. Away from the worries of life nature always waits for you to enjoy. Trees also face the wrath of storms and so many other difficulties. The struggle of man and trees is the same but great grief and tragedy only falls in share of mankind." the moderator said.

"I saw life enjoying the layers and the depths of the sea, marking its presence to mountains, playing in every corner of the earth and making nature beautiful. It was spreading its fragrance on the earth, in the sea and in the every part of the universe. Nature was bestowing its blessings everywhere. When I saw the routine under the burden of life, I also started to enjoy the feeling that I am also toiling with other creations and creatures." the moderator continued but the multicolor soul did not respond.

"Once I looked towards nature and forgot myself. Nature was caressing me everywhere, in every corner, at every moments, all around there was nature and nature. I, the acrobat of life, was busy demonstrating my own acrobatics. When we, nature and I met, there was no way to separate us." The moderator disappeared after speaking, without seeking a response from the multicolor soul.

The multicolor soul was transcendent of all these views and still watching life very closely. It was enjoying every moment of life. The multicolor soul was still thinking about its other companion souls. It noted some changes and started carefully watching those changes.

_4_

"There is nothing like north, south, west or east in the universe; it's the eyes of life that make directions. Then, why I am feeling that I am on the north end?" The red soul pondered.

"Maybe it is due to my last wishes when I was in a man's body." He continued to muse.

"May the light that separated us put us on the earth, and it may be the north end of earth." It was red soul speaking.

"I can imagine the value of life in the beautiful, but tough snow-covered places. I think life starts from here. I can say the color of life is white like snow. Below the snow color, there is red like blood. I can see life wandering hither and thither. Maybe I can't understand the real value of life, because I am looking at the value of life from my own point of view. Actually, life is more valuable than in my imagination. In the real sense, the earth and water are the roots of life, which gets air and light from the sky. Nature gives full freedom to life, but unkind wishes may make life difficult. The relationship of life to nature is like the relationship with one's parents, who gives birth to new life. Life looks beautiful in the lap of nature, but worst in our mean somas. Life looks so beautiful when it passes through the ways of nature. Life is a beautiful flower in the hands of nature. It's the beautiful moment for me to enjoy with life and nature." The red soul continued its thoughts.

"It's good time for me to see life. It's a good place to see life. It's a good atmosphere in which to see life. Nature is spreading its goodness all around over the earth, and there is life to see." The red soul felt comfortable and, so far, was forgetting the pain of separation from the companion souls.

"Life is so beautiful—I have never seen it like this before. It's actually a beautiful thing, but some of the sufferings of life making it so tough. That reality is the truth of life. Life is not the name of any easy task. To understand nature and its task is also not easy. We can't understand the task of nature in our mean somas. To understand it, one has to rise above, leaving meanness behind." The red soul enjoyed the moments of nature and life.

"Life bears the destiny of nature, as well as danger from life to life. It looks as if life is a great struggle. Struggle makes life more powerful. Struggle is the only way to keep the life moving towards the destination. It looks seems everywhere that life is about doing struggle. How brave life looks, brave in doing struggle in the lap of nature, through the natural ways. Struggle is the beauty of life. Truth and honesty are the most important pillars of the struggle of life. Life looks beautiful, when struggling to live on the earth respectfully." The red soul continued to observe life.

"Shine and shade of life, day and night of life, light and dark of life, including good and bad of life, all is going through time with the important fact of life and death, whose account may be in the folding and layers of time. It is difficult to rummaging through or around the folding and layers of time as nobody knows the length and width of time and sky. Only the presence of life is important, or in other words, you may say the existence of life is only important on earth." The red soul was enjoying the present moments.

"Nobody knows whether life is made for nature or nature is made for life. How life comes on earth, when it comes on earth, in what shape it comes on earth. How life developed on earth. How only man became so powerful. Why the form of man became so powerful. What is the role of life in nature? What is role of nature in live? What is the importance of any one towards another?" The red soul said.

"Who became the materialistic and who remains living in the lap of nature, by abiding the rules and the law of nature? What the material world has given to life and what freedom it has snatched are not difficult questions to answer. Life has the same value without the material world, but has no value without nature. Only nature is the powerful source of life. Nature is not only the material world—it is above that. It is foolishness to say that nature is a material for life." It was red soul's view.

"The material world is nothing except selfishness. Selfishness gave birth to the material world. Material is things collected from the resources of nature. Some understanding of nature is not enough to say that nature is a material. The natural world has not yet been finished. Actually, the material world has no value without nature. The material world still depends upon nature. Without nature, the selfish material world is nothing. Life still takes breath in the lap of nature and fully believes nature. Material is not the source of breath of life." The red soul continues expressing his views.

"Nature welcomes life everywhere. Earth, which nourishes life; water, which irrigates life; air, which gives the breath to life; fire, which gives light, path and heat to life; sky, which gives space to life' earth, air, light, water and sky, maintaining their control, all are appreciating life and life is enjoying the blessings of nature," the red soul saw.

"Nature keeps busy fulfilling the needs of life everywhere, and life keeps busy accepting all the blessings of nature," the red soul saw.

"There are songs of nature everywhere for life. There is music of nature all around for life, and life is dancing everywhere in nature. Both nature and life are looking very happy, creating this universe," the red soul saw.

"Nature is giving birth to life everywhere like a mother, and taking it in her lap after death. It gives birth again and takes in her lap after death, and this unfinished sequence is going on everywhere in nature," The red soul observed.

"I am enjoying being at the north end, watching life; still, I have to see the different sides to find my companions, who are on the same mission as I am," the red soul said.

"The rivers are taking shape from snow melting with the sun's warmth. Nature is showing also its other facets, with the beauty of mountains, trees, flowers and so many other colors of life. What a beautiful scene it is." The red soul was enjoying the interplay of life and nature.

"Days and nights, winters, summers, autumns and springs are coming and going. There is life in between living and the passing of its time. Life is walking, running, creeping, swimming and flying in nature," the red soul saw.

"Life is enjoying the songs of nature, music of nature, fruits and gifts of nature, food and breath, and also bearing the storms, tornados, gales, floods, hurricanes, tsunamis and earthquakes of nature." The red soul saw all this.

"Life is roaring, crying, weeping, laughing, enjoying, hurting, suffering, chasing, saving, playing, dancing, singing, living and breathing in the lap of nature," the red soul saw this, too.

"Nature is made for life, and life is made for nature, and both are playing their roles according to plan, coolly and calmly. Animals are fighting for their rights and birds are struggling for their rights, along with life which is also struggling everywhere on the earth. Nature is blessing life everywhere including man, but man is busy in doing his best to win over nature. Man's struggle to control and possess nature is becoming the worst in the universe. Man can't beat the nature. Man is nothing without nature. Man's struggle is useless, because his life is the gift of nature and is nothing without the fruit of nature. Man's struggle is not for humankind or creature, or for nature," the red soul opined.

Life was looking very busy, as well, as the red soul journeyed from the north end towards the south.

The red soul saw, "Human beings have gained more than enough knowledge everywhere and are craving the fruit of that knowledge. They think that knowledge is the only route to the happiness of life. There is nothing in them to take the fruit of nature. There is nothing in them to take the freedom in the lap of nature. They have forgotten how to enjoy nature and indulge themselves in the things developed by trade, for the trade and of the trade, from the resources of nature, which are common to all creations. Human beings struggle to get the few things developed by trade to live, as they are thinking the struggle of life is only to obtain things made by man for trade. They have forgotten how to struggle to obtain the real fruits of nature. It looks as if they have forgotten the real meaning of life in the lap of nature, so they don't want to struggle for that. They are mad to grab man-made things."

The red soul saw, "While other creatures are busy enjoying nature and living in the lap of nature, human beings are hiding under the roofs of their houses, which were created out of their knowledge. They have no willpower to overcome this crisis, because their knowledge has declared the house a luxury that must be sold at their own expense, and some have died to obtain this luxury."

The red soul saw, "The human being is busy fulfilling his false wishes with the help of the things he created things by thinking, and with the worries of the future that are standing on his past, and by hiding himself and being afraid to face the present."

The red soul remarked, "The human being is praying for an unforeseen heaven after death, without any effort or struggle in life to enjoy the presence and existence of the heaven of nature on earth."

The red soul saw, "Other creations are not behaving like the human being. Instead, they are worshipping the fruit of nature and believe nature, but the human being is trying everywhere to dominate nature and to declare himself the supernatural. While the whole of creatures dome depends upon nature to fulfill their daily needs, the human being is busy stockpiling the things that others need and preventing others from using them."

"Man is trying to possess and control nature in his own ways by discriminating, not only against all the other creations, but also against fellow human beings. Human being is more affected by the so-called progress of man than all the other creations and nature. Human beings are suffering more from the crisis of the so-called progress of man," the red soul saw.

"Except for a few, the human being's condition everywhere is more pitiable. Man has not caused danger to the whole of nature and all other creatures but human beings themselves have become ill with acts of progress. The human being is tightly bound to the powerful system that supports the possession of few men over the many. The resources of nature are being captured, possessed and controlled by few men and the rest of the human beings look towards them as they die waiting for the day to come when they will also get the fruits of progress. But it is all in vain," the red soul saw.

"The power of the system keeps the terror of ammunition in the minds of human beings everywhere to shut their mouths against this unbearable discrimination under the names of laws that favor their ill treatment." The red soul saw and said, "How these powers have made all of nature and creatures a commodity, and how they are selling creatures and nature and its fruit in the open market."

"Man is trying to enslave nature and creations, as he has his own population. Man has taken into his hand all the resources of nature and is crippling nature," the red soul witnessed.

"Man has taken into his possession the natural resources on, in and of the earth at large scale, and is separating it from earth at a speed at which nature is unable to renew them. The man is wasting these resources to fulfill his pointless wishes, which have no use in the future, except to make the human being more ill and sick." The red soul saw all of this.

"Man over all the earth is living in fear created by the progress of man. The other creations are also living in fear created by man. Everywhere there is fear of the progress of man. Somewhere, man intentionally and deliberately created the fear in other human beings. Perhaps man is unable to create fear in nature, but he has compelled nature to be wrathful. All the creatures are bound to bear the carelessness and foolishness of man's progress," the red soul saw.

"Somewhere the wrath of water, somewhere the wrath of air, wrath in the sky, somewhere the wrath of earth and somewhere the wrath of fire make pitiable the position of human beings and the other creations, the red soul saw.

"Snow and the glaciers are melting into water, due to the warmth created by the progress of man, and nature is becoming helpless to convert the water again into snow at the speed required to maintain the natural balance," the red soul saw.

"Man's rate of so-called progress is making natural events more unusual and dangerous, not only for the human being, but for the other creations, as well," the red soul saw.

"The science which man developed for the benefit of man is becoming the curse of man and all the creatures. All are bearing this agony." The red soul saw a man crying, "No destruction! No doomsday will ever come. No doomsday. No doomsday."

He had just wakened up from a deep sleep and was still not able to separate himself from the dream he saw in his sleep.

Now he was sharing his dream with his friends, due to his blind faith that if you disclose the dream to others, the dream will never come true, but if a man fails or forgets to disclose the dream, it will always come true in real life.

He relates, "When I met the ocean for the first time, the ocean overwhelmingly welcomed me by touching me through its waves. I was thinking, the ocean has come to greet me, even though it was its habit and it welcomes everyone in the same way, but my feelings were out of control with happiness because of my warm greeting. I was trying to keep these moments of feeling in my memories for a long time. My sequence of emotions and feeling was suddenly broken when the ocean suddenly started addressing me. Maybe these were the answers to questions buried in my mind since long ago."

The ocean was saying, "Who says that when man discloses all the secrets of nature, then there will be the final destruction of the world? This is not true, at all. They are ignorant about the reality or concealing the reality. I will tell you the reality and truth about what is happening and what will happen."

I was shivering with fear and sweating profusely. I was not feeling myself able to accept any truth at that moment. I cried, "Oh, my lord! Mercy! Please, mercy!"

The ocean ignored my fear and cry and continued to say, "Don't be afraid, my son. Be brave and accept the truth. The only way to get relief from fear is to face the truth. Nature has hidden nothing from man and is like an open book before man. What man will discover more than this? Man is ruining humankind under the burden of his discoveries, but nature shall remain open before man. Man has to look towards his feet. Otherwise, he may lose the earth beneath his feet, and will find nothing from the sky. If man went to another planet on tour, he might lose his own planet, due to his own ill wills. Man is destroying and finishing the neutrality and balance of earth with his mal-discoveries to fulfill his useless and mean wishes. Doomsday is not far away, due to man's own wrongdoings."

I was crying in the dream, "No destruction! No doomsday will come." I suddenly woke up from the deep sleep and I am still unable to separate myself from the dream."

The red soul felt happy that man is too much worried about his doings.

The red soul saw a man, serious about the problems of life, asking a philosopher, "You have traveled from the tops of mountains to the sea, north to south east to west and have seen all the colors, concepts, standards and corners of life. So, which journey you think is more difficult?"

The philosopher said, "The journey of food from the hoarders of owners to the stomach of indigents is the most difficult, and nobody is serious about it. This distance and cost is increasing day by day."

The red soul than saw a son was talking to his mother about rights and duties. He was telling her that duties are more important than rights. He was telling her that sometimes we give importance only to right and forget about our duties. He was explaining how changing values are making us ignorant about the realities. He was saying that man has become selfish, ignores others' rights and forgets duties. He was trying to convince his mother of how aware he was.

His mother listened to him seriously. She moved her head up and down as she listened.

After finishing his talk he looked towards his mother, as he was expecting some words of appreciation from her.

The mother looked deeply into his face and said, smilingly, "Oh, my silly son!" And shaking her head in a negative sign, she said, "You must be hungry. I am going to bring some food for you."

The red soul then saw a big tree telling small plants to look up towards his growth and to the sky. A small plant said, "I look up towards sky only when the sun rises in the morning."

The other said, "I look up only when hot rays burn me and become unbearable."

The third said, "I look up only in the moonlight."

The fourth said, "I look up when the morning star shines."

The fifth said that he looks up during the morning dew. A man came there, and the big tree put the same question to the man. The man followed the big tree immediately and looked up towards the tree and said, "You will fulfill my all needs," and started cutting the tree.

The red soul then saw on the bank of a brook sparrows recently returned to the grove of trees for their night stay on the branches of trees, after taking their daily feed away from this, their dwelling place. They were singing harmony. They made two or three bevies around the trees and took their places to stay. The atmosphere was still melodious with their songs and voices.

The red soul saw a nomad coming with his family, snatching the rein of his jenny. Their child was lying in a loose sling of cloth tied around the waist of his wife on the one end, with the rest of the cloth over her right shoulder to the front side tied with the same cloth. Their child was crying with tiredness, hunger and cold.

It seems the people of a new colony had not allowed them to stay in the vacant place of the colony, so he was coming away from there. The nomad looked hither and thither to select the proper place to install his small tent made of old cloths, with the aid of some sticks.

The child, when heard the melodious songs of the bevies of sparrows, stopped weeping. For a moment he forgot his tiredness, hunger and cold.

It looks as if the sparrows must have sung a ballad today, as the red soul saw there were a rage, revolt and ebullience in the eyes of the little child.

The red soul then saw a lion in the cage of a zoo, thinking about his fate. He never ate stale meat when he was living in the jungle, and now he was fully dependent on stale and contaminated meat to keep him alive as a slave, and saw people were laughing at his slavery.

The red soul then saw some people who were saying that, at some time in childhood, the face of the god now carelessly passing through the places of gods strayed in the hovels and huts of indigents, the poor and beggars. Gods were dying from dangerous viruses, hunger and thirst. The great mother earth, which itself was in the possession of some power, was helpless to save them from the crisis.

The red soul saw some unforeseen darkness around itself. It stopped his cause and was now bearing some changes in it.

_5_

"There is no end. But, maybe that is because I am on earth, so I am really on the west end. How beautiful it is. I don't believe in west or east. Where east you think ends, west starts; and where you think west ends, east starts. It's your thinking, where or what you see. The name given by you is your eye's deception. There is no end of east and west. However, I presume I am on the west end," the green soul said.

"I don't challenge any ideology, what it is and why it is happening—how, when or anything other, than to collect any factual position or data. I don't want to understand these things. However, I have promised my other companions to discuss life, so I must discuss to fulfill my promise, even though we are not together now," the green soul expressed itself.

"Life is beautiful to enjoy, but it is tied to the other creatures. Freedom of life is to save yourself, save your rights to keep yourself alive in the nature. From where your rights start, others' rights to defend themselves also start from the same point and at the same time." The green soul continued to express itself.

"If life is to enjoy, then on the other side, life is also the food of other life, and life ends life in a moment," the green soul continued.

"Nature is for all, but nature never enters into life's routine. Life has to defend itself, to do itself, to live itself or to fight for its freedom," the green soul went on to express.

"If the law of the land gives life the right to grab the rights of others, then the law of nature prevents grabbing others' rights. Nature has given to life a small stomach, but unbeatable nature and earth. Life can't conquer nature. It may control the earth to a limited extent, up to it selfishness, to which nature is able to defend itself or respond." The green soul expressed these thoughts.

"Life is not dependant on life, but it depends on nature. Without nature, there is no meaning of life. If nature exists, then there is life, too. Life has the same value without the kingdom and discovery of life, but has no value without nature," the green soul explained further.

"Life comes on earth, enjoys the fruit of nature and goes away. How many lives came on the earth, enjoyed nature and expired, nobody knows. Nobody knows how many lives are still on the earth, living in the lap of nature, and when they will be finished. Nobody knows how many lives will come on earth and enjoy the fruit of nature and be finished. Nobody knows how many lives take birth on earth and how many lives leave their somas and become part of earth every day or every night, or in a moment. It's foolishness to judge how nature works. Nobody can judge nature's workings. No one can even praise nature. How many lives or births are needed to praise nature? It's beyond anybody's expectation. Life takes birth and goes away in praise of nature, but remains unable to praise nature," the green soul said.

"Who thinks he can seek happiness with the worldly things made by him from natural resources is not only wrong, but lives in fallacy. Nature is truth. There is no reply to nature from life. Nature is the answer of life. Nature nourishes life. Nature is the witness of life after death," the green soul said.

"Ideologies, doctrines, canons, laws and rules of life may be the beautiful colors of life, but are not permanent. Above all, nature is permanent for life. All these change with the passage of time, but nature remains there in the same way, serving life," the green soul observed.

"Nature always fulfills the needs of life. Nature has everything to provide life. Life is the ornament of nature, but life has nothing to give nature." This was the green soul's view.

"Life should always be thankful to nature, which nourishes life without expecting anything in return. This process has continued for nobody knows how long, and nobody can presume how long it will continue to serve," the green soul said.

"I am on the west end and have to move towards east to find my other companion souls," the green soul thought.

"The sun rises in the east and sets in the west, distributing new power to life every day. Life may not remember the reward, due to his routines and being busy with his selfish wishes, but nature expects nothing for its services rendered to life," the green soul expressed.

"The creature that was making nature beautiful is enjoying the boon and blessings of nature everywhere. Every day is beautiful with nature, every night gives rest to the creature, every season makes the creature and nature more beautiful and this circle shows the unbreakable relationship of life and nature." The green soul expressed more.

"It's true that the earth revolves around the sun with the other planets, and rotates around its axis and makes day and night and seasons, but my journey from west to east sees the sun rising and spreading its light and shine over the earth, which is distributing combined happiness and sorrows to life," the green soul said.

"Maybe there is change in living, maybe some needs differ, maybe some standards differ, maybe some atmosphere differs, maybe some weather differs, but the basic needs are the same of life everywhere, and the ways to fulfill it are the same." the green soul saw.

"The color of the happiness and sorrows, the blood and the life is the same everywhere in the world," the green soul saw.

"Mountains, rivers, bays, oceans, plains, air, water, fire, sky and earth and life over it, are the same everywhere," the green soul saw.

"I can see the science and discoveries of man everywhere on the earth, which are polluting nature and causing so many difficulties to nature, creatures and even to humankind. He is doing all this under the guise of benefiting human beings, but destroying the natural resources for nothing, with no benefit, except to create more and more difficulties for human beings and the other creatures," the green soul said.

"The science of man and his discoveries are not only destroying the beauty of nature, but also creating so many dangers to nature, creatures and even to the human being. Smoke from fires which fill the air after so many killings of men, is prevailing everywhere on the earth and becoming more and more dangerous day by day. It's wrong to think that fire can be extinguished by fire. This is a wrong concept. Imagine if the world were without ammunition," the green soul said.

"Man put me during his life, and now the generation following him, in the list of enemy birds of man, only because crops sown by man were their favorite food. What a cruelty and enmity shown by man." The green soul expressed.

65

"Man put the birds in the list of friend birds and enemy birds, according to his personal needs and losses. He extinguished the enemies and favored the friends. Man's behavior towards nature and creation is very rude, due to his selfishness," the green soul saw.

"Many other birds are the victims of man's cruel behavior. Some of them lost their dwelling places, some lost their trees, some lost the right to water and pure air, some of them are suffering because of pollution of air, water and sound and some lost the right to food. Some birds are bearing the cruelty of man in every other way to fulfill the wishes of man," the green soul saw.

The green soul observed, "The progress of man has created pollution to make the roads on which the progress of man runs very fast and creates a high level of noise that can cause his heart failure, deafness, mental disorders and so many other serious and non-curable diseases. Who cares for the other creatures, if he is unable to save his own human beings from the noise of his progress? People who have already made their houses near that progress of man are now the victims of his progress and are unable to save themselves by flying away like birds. Man is bound to hear and bear the so-called progress of man. There is no solution left in the hands of man to save himself from his own progress."

The green soul saw, "Man has nothing to give his coming generation to live calmly, except to bear the pollution of noise caused by his progress."

The green soul saw, "Some men are hiding themselves from the noise of progress in mountains and other safe places, but are unable to get away from the progress."

The green soul saw, "Man is not fair even towards human beings; how can other creatures expect anything better from man? There is nothing strange if birds, animals and other creatures are suffering due to the pollution of man's progress."

"The birds can fly and are able to go away from this pollution. They are able to save themselves, as they have not collected too much material for life and the future, and they do not have big houses like man's to carry with them. Nor do they need too much food to store, as they fully depend upon nature. They have no worries about the future, and they have no problem to fly away from this pollution," the green soul noticed.

"The animals and other creatures have to suffer the crisis like other men, even though their needs are not too much like men's. They are bound to bear this pollution, as they are also unable to run away from this crisis," the green soul observed.

"There is no end to this pollution of noise, and it is likely to increase in the future, as the increase in progress is likely. Man, his coming generation, nature and creatures will continue to suffer this pollution and progress of man," The green soul said.

"The progress of man not only created this nuisance, but also to fulfill his mean wishes, he digs the earth and takes out gases, metals, minerals, elements, coals, stones, oil, liquids and fluids on a large scale. He destroys the natural resources to fulfill his mean wishes and throws the polluted and wasted things on the earth, air and water. Man is doing nothing to save the earth, air and water, which is making nature helpless to serve all creatures, and this is happening only due to the selfishness of man." The green soul watched.

"On the earth, man is using mercilessly the things that are part of the earth and nature, to fulfill his useless wishes and needs created by him to show the miracles of his progress. These things

68

he has collected from the resources of earth were created to maintain the balance of the earth in the planet. Man is not only making nature handicapped and polluted, but is directly intervening in the routine course of nature that nourishes all creatures," the green soul remarked.

"Man's progress is not only to fulfill his needs, but is also to support his useless ego. The progress is not making life easier, but making it more difficult than ever it was, and putting all of nature and all the creatures in crisis," the green soul saw.

"The crisis is not going to be solved in any way, but is becoming more difficult to solve than ever," the green soul saw.

"There is no part of earth in the east, west, north, or south that is not affected by the progress of man. It is useless for me to say that I am going to the east from the west. Man has joined hands everywhere in making this so-called progress and thinking about it has become same. It's man who is responsible for the said progress over the earth, but the consequences of the progress are borne by the whole of nature and all the creatures," the green soul saw.

"Nature and creatures over the earth were busy in their routines in spite of man's acts to harm the beauty of nature and to damage its strength," the green soul saw.

"The mountains with their beauty of trees and many other useful things, some covered and some uncovered with snow, flowing rivers, brooks, fountains, waterfalls and streams from mountains down to the sea, oceans and over the earth—the whole nature and all creation were busy at their work, freely, without any hesitation or fear of the past, present or future," the green soul saw.

"The work of nature and creatures was so great, even beyond the thinking of man or its peak progress in his mind, that no one on the earth can match the same. Even the pride of man is nothing before the nature," the green soul observed.

"The cause of air, water, sun, sky and earth was bigger than the cause of anyone. Man's progress is his selfishness, but there is nothing in the cause of nature like selfishness." The green soul watched.

"Everyone on the earth, including the pride of man is under obligation to nature, and the blessings of nature to its creation are without any selfishness," the green soul saw.

"Everything over the earth was playing its role towards others on the earth with the blessings of nature," the green soul saw.

"There was no discrimination in distribution of the blessings to the creatures. Blessings of nature were open and for everyone, except the control and possession of man," the green soul saw.

The green soul saw a swan and a man were talking. The man looks like a saint was telling the devotion of man towards god and nature. He was saying the man a devotee of nature and god.

The swan pointed out the men who don't believe in god and who are saying that nature is a material.

The saint tried to count the sacrifices of man towards the nature and human being.

The swan said, "The sacrifices of man for human beings are only for the human beings. Even human beings are not fully satisfied with their sacrifices. Human beings are ruthlessly killing human beings in the name of their sacrifices, who sacrificed their lives against this roguery. They have been given the path only when they were alive. After their death the path falls into the hands of those persons who do not have the same spirit, experience and knowledge of life, and who have no control over the situation. Moreover, always this struggle continues between the rich and poor. The trader makes the sacrifice a miracle, instead of relating it as a struggle, and mixes in the routine problems by dividing the ways of life and belief. The ways of theists and atheists are the same to the poor, and thinking is to save the rights of poor human beings from the brutality of richness. The believers cannot understand the deep meaning of this philosophy. It is not possible that all the time everyone will work with the same faith and spirit. The result is that the same human beings remain divided into isms, ideologies and faiths and ruthlessly killing each other for the victory of their faith, which was not even in the mind of the originator of that theory. The rich traders always played their role in indulging the people in this crisis. The philosophy of human beings to save the rights of poor common people from the brutality of richness remains absent."

The saint agreed with the swan and said, "One has to come forward to bring the truth out from the darkness. Man's created power is nothing before the power of nature. Man is running to gain the power created by man. Man's race to gain this power is nothing and is mortal. The role of nature is more important. Even if man, with his own pride, destroys the existence of earth, the role of nature shall not be limited. In the sky of planets, this happens all the time."

The swan said, "You can see the history of man's progress stands on the pillars of his knowledge. Man was living like other animals on the earth in the natural way. He started the era of progress for the welfare of and to facilitate humanity. He assisted man with progress but started to capture the resources of nature, preventing other men from possessing them by making into commodities. Man divided the earth into countries by its mean wishes to control the poor and to spread its trade. Trade invaded other countries to spread his control over the earth. Trade killed human beings mercilessly in the face of false and fake ideologies ignoring the laws of nature."

The saint said, "You are right, but always in this struggle most human beings stand against this and continue to struggle against this brutality. Traders have made strong systems in their favor, in which man is compelled to think like this; he is too weak to break this brutality. He is bound to think that he is unable to spread the law of nature instead of the law of man. But this is also a reality, that most human beings like the law of nature more than the law of man. They feel that nature is not material as some of the human beings think. Nature is bigger than the thinking of man. It may be true that man is using nature as a material, but that is because of his foolishness. Most human beings are bound to live in these systems, and they have no other option, but that does not mean the struggle for victory over the natural atmosphere has been finished forever, and man's efforts and devotion towards nature has been ended forever."

The swan said, "Theist and atheists are throwing their theories around, but the killing over the earth to strengthen the boundaries of the countries to secure trade within and across the boundaries still continues. More deadly weapons are being prepared to secure the trade. The life of all human beings, and even all the other creatures, is at great risk with man's ideas for trade and ignoring the laws of nature."

The swan saw another man putting his fingers on the trigger of his rifle, aiming at her, and flew away before becoming his victim.

The man shut his eyes, looking as if he were a great devotee of nature and praying.

The green soul then saw a young pair of sparrows, recently married, talking with each other. The male sparrow was raised in the jungle and had never seen man's progress. The female sparrow was raised in a city. The female sparrow now brought him to her nest in the city and was telling the stories of man's progress.

When the female sparrow made a dropping, the male sparrow started taunting her. The female sparrow, to distract him, said, "Man has made so much progress that he has made such things in their homes to answer the call of nature at home."

"He must have called the dangerous diseases at home," the male sparrow said.

"No, he has made system to clean everything," the female sparrow said.

"Man cannot make a system to clean the air. Only nature has this system. His system might make him sick," the male sparrow said.

"Man has developed a system to cure all diseases. He is able to change the organs of man by operations," the female sparrow said.

"Now he will trade the organs of his poor men," the male sparrow said.

They heard the call of their elders and commanders and they fled away to join the bevies. They were leaving this place for another place, due to the change of season and to find big treasures of food. They were bound to abide by the rule of their elders and commanders. Due to the rule, they have to separate during flying. The female sparrow fled away to one place on earth, and the male to a different place. After a few months, when they returned, they met again and welcomed each other warmly. They decided to live together again, and they also decided to live in the deep jungle. At night they discussed their experiences during their journey and stay abroad.

The female said, "I saw that people were fighting for the birth place of their prophet. I don't know how many people died in the fight because that was also the religious place of another community. I don't know why they were fighting, when their prophet left everything for their principles. I think they were catching that hand that is signaling the way to the destination, instead of going on that way."

"True. I also saw the same thing where I was staying. They were also fighting for the birthplaces of their gods. And they had been fighting for many centuries about the same issue. Their prophets gave their lives peacefully, without any violence and voluntarily, for the principles against the trader world's cruel behavior against common people, but now common people are fighting and traders are doing trade in their blood. I think their gods were born at one place and chose to go into the open world, but people are going from the world to their birthplaces, instead of following in their path. People, I think, don't want to see the whole world like their gods, but want to die only for one place where they born. What a funny thing," the male sparrow said.

The female next day expressed her wish to see her old city. They flew away to the old city. When they were sitting on a tree, they saw a dog playing with its owner. After some time, the owner went inside the house. They were happy to see the activities, happiness and love of dog. The dog barked at them. When asked the reason, the dog said that you are coming to see my pitiful life. They expressed ignorance about his pitiful life. The dog said that his owner castrated him, and now he can never have sex. He is just a toy now, a living dead thing.

The dog further reported that man's life have become a trade. Some are selling today for a better tomorrow. Every time, tomorrow comes with more darkness. So man's behavior towards us became a trade to prevent others from doing trade. We are suffering the cruelty of man.

The green soul then started once more its journey towards the east from the west. However, darkness was prevailing around her and it became silent.

_6_

"I am starting to go towards north, between the east and west, to find my companions, who may be looking for me, too. There is no difference of life in south, north, east and west, but I will try to fulfill the promise I have made to my other companion souls.

"I know the fate of life; maybe that experience is from my own knowledge of life. Dogs are generally pet animals. I do not know the history of how dogs became the pets. Maybe the faithfulness of dogs towards men made them pets. Faithful dogs now are bound to live under the laws of man. Man's law is to kill dogs that are not pets, calling them strays. Stray dogs he thinks dangerous to human beings. However, it is good news for stray men, who do not face the threat to their lives that stray dogs face," the blue soul remarked.

"Now dogs are treated as the personal property. Man may have control over his children or not, but he binds to his control the dog kept by him. After eighteen years, the child becomes free to make his own decisions, but the dog has no right to live without his owner during his life," the blue soul observed.

"Dogs are not independent now. They are bound to live under one owner, to live according to his wishes, to follow his wishes, to follow his discipline, to follow his orders, to follow his habits, to follow his food and living standard and to follow everything he wants. There is no way for dogs to get out from the clutches of his owner," The blue soul said.

"Dogs no longer know the meaning of independence. They are the pets, and under the laws of man they remain a pet of man. There is no independent identity or wish of dogs under the strict laws of man," the blue soul perceived.

"Man has the boundaries of his country within which he is independent, but the dog has only the boundary of the house, from which he is not able to come out, and there is every arrangement for him inside the boundary of the house of his owner," the blue soul expressed.

"Dogs have no freedom to choose even their owners. They are bound to live under the ownership of that man who has given the price fixed by the owner of the mother of the dog. The mother has no right, except to give birth, to keep him with her according to her wish. The mother has no right to choose the owner of her whelps or pups and no right to take any decision for them or for their future," The blue soul commented.

"There is no other example in the whole world of slavery to compare with that of dogs. Dogs are bound to bear the roguishness of man; otherwise, they should be ready to die as stray dogs," the blue soul said.

"Faithfulness is also the crisis in this world. All this happened because dogs showed their full faithfulness towards men. Men recognized their faithfulness, but they recognized it only for them. Dogs have to live as faithful, and no independence is permitted beyond this faithfulness," the blue soul observed.

"It was the weakness of dogs to accept the slavery of man. But this happened not only with dogs. This happened everywhere, with all. So many animals and birds are facing the same situation, due to their inability to free themselves from the grip of man," the blue soul saw.

"All big animals have no independent place on earth to live. Everywhere man has stamped his ownership. Almost all the animals are bound to live at the mercy of man," the blue soul saw.

"Animals are more affected by the progress of man and by his taking the earth into his possession. The lust of man to take the land into his possession has made the life of animals pitiable. It has become difficult for animals to live on earth at their own will. They are at the mercy of man. Their life is fully controlled by man and depends upon the mercy of man," the blue soul saw.

"Birds are also affected by the progress of man on a large scale, but their condition is not made so much pitiable as the animals'. Animals are bearing the cruelty of man more than the birds," the blue soul saw.

"Animals have lost all their rights. Most have lost their every right over the earth, even the right to enjoy sex, food and life. They have become fully dependant on the wishes of man, for what he provides or what he not provides." The blue soul saw the state of the animals.

"The meaning of slavery can be observed from the slavery of animals which cannot get freedom forever from the lust of man's possession over the land," the blue soul observed.

"Perhaps some animals that can run fast, those with small bodies, easy to hide, are able to live freely to some extent, but the lives of big animals and those that cannot save themselves from the clutches of man's egocentric desire to control them and store them for their use in the future, are in a pitiable condition," the blue soul saw.

"Nature is so beautiful, but man has made this his personal property and treats nature as if it were part of his trade, preventing other creatures from using it, as he is preventing the poor men from using it through its trade," the blue soul observed.

"Is man a social creature? Can man's progress be considered as a symbol of his sociality? In reality, man's progress has destroyed the meanings of his sociality. Man himself has become dangerous to his own society. The persons who made the society beautiful with their blood, and gave meaning to their strong stance against those who made the society dirty by their trade and who were flooding the blood of human beings, are becoming the victims of the same thinking, and society is again in the grip of the same brutality," the blue soul said.

"Creation of the society of nature has needs different than the society of man. Man's society has been developed with the knowledge of man, which does not need ways that lead to nature. He is in the cage of his own ways, which lead to destruction," the blue soul said.

"Man created society to save itself from crisis in difficult times, which he bears alone and to strengthen the unity of man against the difficulties he bears as a whole, but the result will be so terrible and dreadful, as even man should have not presumed in his dreams," the blue soul said.

"Man made society, which the powers shaped into the countries, and countries under the names of developed, developing and undeveloped countries. People became rich, poor and indigent. Money developed by man has gone into the hand of capitalists and traders. Other people became the victims of it. If any of the poor raises an objection to it, he has to bear the atrocity and brutality of the system developed to be the safe haven of capitalists and traders. Sometimes, one raises the voice against this system has to hang on the cross, to be hanged, to be executed or to face dreadful consequences," the blue soul said.

"Man is bound to live, work and obey the dictated terms of the system, and is unable to ever escape from this dreadful claw," the blue soul said.

"Man shows his faith in god and some say there is no god. Every time, he forgets that his existence without nature is not possible. If a man has not faith in the work of nature and the laws of nature, then how can he be truthful, and how can he follow the principles of nature and life? Truthful life is only to obey the principles and laws of nature. Otherwise, he cannot be the follower of god, even," the blue soul said.

"No ideology can be above the laws of nature. The philosophy of nature is above all. No one can be great without following the philosophy of nature. If ones adopt the philosophy of nature, he needs no more than this. All is in nature," the blue soul said.

"The life of other creations is not confused like life of man as they believe and depend upon nature only. They don't fall into the dispute to be atheist or theist. They enjoy more freedom. They enjoy a more truthful life with the blessings of nature than man does. The needs of life start with nature and end with nature. Man doesn't want to understand that he takes birth on the earth with the blessings of nature; his life depends upon nature, and after death, becomes part of nature. But man, during his life tries to forget all the blessings of nature and always goes against the principles of nature, trying to depend upon the things made and developed by man. Man becomes confused among his worldly developed things and becomes ignorant about life-providing nature," the blue soul said.

"The great tragedy of man is his mistaken belief that his progress is everything. Actually, the things made, prepared, discovered, grown by man may have given nature another shape in his trade, but are not of much importance in the course of nature. The course of nature is entirely different from the course of man's progress. The natural course is without any selfishness, but man's progress course is full of selfishness. The course of this man's progress has become the crisis and hurdle to enjoy the natural course as freely as do so many other creatures living on the earth," the blue soul said.

"All of creation enjoys life more than man. Man bears excess sufferings due to this crisis in his life, more than the other creatures. The life of man is more pitiful than the other creatures'. Man's life is more difficult than the other creatures'. The creature bears sufferings of this crisis of man's progress and due to man's doing, but not due to nature or its process," the blue soul said.

"Man cannot live on earth even for one second without the blessings of nature, but is unable to understand this and tries to say that his doings are more important than nature, and that man depends upon his blessings instead of nature's," the blue soul said.

"It looks as if the progress of man has broken the back of the society and he has grabbed the whole power in his hand, and is grinding the people in the quern of progress and trade," the blue soul said.

"You can see the dance of poverty everywhere on the tips of power, and the needlessly flowing blood of the poor here and there, and the poor tied tightly in the laws and flaws of progress of power, and his cries have no value at all," the blue soul said.

The blue soul saw some pages of unappreciative history flying hither and thither.

"History is a painful memory of life in the past, from which you analyze your present to bring about a better future, but it does not happen in history," the blue soul said.

"It was the history of the bull, under so many names like ox, bullock, stud bull or steer. It was the history from the age when man was living on the earth with them, enjoying equal rights; but sometimes man hunted them, only for meat. They were living in equality, as they had right to defend and save themselves," the blue soul saw.

"Time passed, and man started acting with his clever mind to capture and possess land all over the earth. Bovines became the first and the immediate victims of humans. The milk of their cows became a favorite source to maintain the stamina of their bodies, and meat became another favorite source of food. More than this, man used the strength of their race for cultivation purposes," the blue soul saw.

"The first cruel attack on the bulls was when man thought that sex was the source of one's power. To keep the bulls healthy, they started to castrate the bulls to use them in cultivation and other purposes. Now their birth over the earth became only to serve man, even in most cases, without the right of sex. Whenever they became slow or tried to stop working, the man gave a blow of the sharp point of his stick on his testicles to show the bull the meaning of his slavery," the blue soul saw.

"How many years it served humanity like this, the pages of history did not tell. Current pages of history show the progress of his life. His sexual life has not improved, so far. But his semen is available, stored in the fridges of man. It can be used anywhere in the world without him, and is still also of very high quality, able to produce a top-quality race," the blue soul saw.

"A bullock was waiting for his turn to go to the machine where his head would be cut off, in the renowned company that is very famous for producing high-quality beef in the world. The bull was watching deeply the neglected pages of history. He was thinking before going to his fate about which age of the era was golden for him. History does not give relief to anybody, but he was searching the golden age from the pages of history," the blue soul saw.

The blue soul then saw some political pages of the past and present that are not likely to be changed in the future.

"The wolf this time has changed his strategy, due to an old stigma of dishonesty from a time when he said to the lamb that you are making the water dirty, and the reply of the lamb was that water is coming from your side towards me.

"Again, when he said to lamb 'You spoke to me last year,' the lamb politely replied that he had not yet been born last year. When the wolf had eaten him, he said that that must have been your father." The blue soul witnessed this scene.

"Due to that disrespect in the history book, the wolf changed his strategy this time. Now he was saying that he is disarming the lambs for the prosperity of democracy, as their horns have become apparently dangerous to life in the democratic sphere. Those who oppose the wolf's policies are saying that he is fulfilling his secret mission, but wolf says that it is his secret mission, but it is open in the world for all," the blue soul saw.

"However, this time the wolf has not taken the risk of giving the lambs time to speak. The bloodshed shows the wolf must succeed in his target, and the lambs without horns are looking at the extent of his hunger," the blue soul saw.

"Some pages of science were trying to lasso the sky to bring it down to use according to their wishes, but sky was not coming into their grip. It was becoming deeper and deeper, and moving away from their clutches. Science is trying to take charge of nature, but nature's own wrath washes away all his desire to control nature. Instead of feeling ashamed, science starts barking again after some time, when the common people have forgotten the injuries of wrath," the blue soul saw.

"Man was whooping that he has won everything from nature and is going to win the remainder with his science, but he looks insignificant when he displays the inability to create light like the sun, or even the moonlight, which the moon borrows from the sun and throws on earth and makes the earth beautiful, or the light of stars, which makes the night beautiful over the earth. The light made by man could not enlighten the earth like sun, moon and stars. Man is even nothing before the air given by nature. Man is not able to create air like this. He can alter the balance of nature with bombs, but he cannot create air. In the same way, he is unable to create water, but can throw it out of balance it or control it, to prevent others from using it," the blue soul said.

"Man is not able even to make his muscles or flesh. He cannot make his own blood and bones," the blue soul said.

"Some pages of philosophy were lying in the garbage smeared with sauce and leftover food. Man, bowed with the burden of his progress, was showing himself the great devotee of the philosophy," the blue soul saw.

"People were impatiently rioting to grab the worldly things, but the stick and bullet of law was a big obstruction in their way to achieving their wishes without bowing before progress, and they were bound to say it was the wish of god. The ass of society is going towards the way of destruction with the sticks of traders and capitalists," the blue soul saw.

"Man has shut down all his to doors of freedom and confined himself within the walls of big buildings and boundaries of countries, in which only few are safe and the rest are sufferers," the blue soul said.

"Man is making and developing those things that are not useful for life and losing that which is useful for life. All ways of man's progress are approaching destruction," the blue soul said.

A dog was indulging with his owner. Actually, he was a stray dog, but a beggar started to give some food got by begging. He believed him his owner and remained with him at mealtime. Both were happy with each other, and they never interfered in each other's routine affairs and were free the rest of the time except for meals.

One day, when the man tried to persuade him that he expected nothing from him, but gave him food without any vested interest, the dog started barking at him.

The dog said, "Man made a system in which he has fettered every man and nobody is able to go beyond his bounds, and the key of that system is in the hands of traders and capitalists. They rigged the system like this: they know the bio-data and history of everyone and nobody is able to escape from that. You are bound to beg, and you can do nothing. In the same way, the poor are bound to live in poverty and can do nothing. If man's fate is like this, then what can I hope from your society? You give me some food only to satisfy your ego that you are a very merciful man, by ignoring your own destiny.

Two birds were talking, "Yes, this true, man is using water to mix everything that he has separated from the earth with science. There is no source in water to cure man's blunders; only earth can do this, but who will tell him that you will die, but the knowledge collected by you will cause suffering to all human beings and all things on the earth."

The blue soul ignored them and tried to go forward toward its task, but felt for some strange reason that it could not find its other companions.

-7-

"Everybody looks towards the east, but I am in the east starting my journey towards the west. I pray for the sun to give me strength," the yellow soul said.

"The sun among the planets is so powerful, around which the earth, full of life, revolves. It takes its turns to keep life warmth and healthy by rotating around its axis, makes days and nights and different seasons, spreading its beauty of nature. The earth is the main source of life, which nourishes life with the help of air, sun, sky and water. When I am starting my journey from the east towards the west, the well wishes of the sun shall always bless me with its rays," the yellow soul said.

"The sun in the morning rises and starts its journey from the east to the west by spreading its rays and shines over the earth. May be it is earth's movements that welcome the sun in the east, but the faith is also the same that the sun rise in the east and nobody can ignore this reality. The sun rises in the east daily and sets in the west," the yellow soul said.

"I am starting my journey from the east towards the west. May be my cause is tied to my own wishes, but the blessing of the sunshine shall remain with me," the yellow soul said.

"Everywhere on earth life starts its journey with the blessings not only of the sun, but with the blessings of water, air, sky and earth, without taking into consideration whether they are in the east, west, north or south. Blessings remain with life everywhere and in the same state," the yellow soul said.

"Life is thankful to nature for its blessings forever of his life, as life takes birth in nature till it ends, until the body of life is absorbed back into it." The yellow soul said.

"Life was enjoying nature everywhere and was happy with the blessings of nature," the yellow saw.

"Water, air, sun, sky, and earth were looking busy in nourishing life everywhere," the yellow soul saw.

"Life everywhere was busy in accepting the blessings of nature," the yellow soul saw.

"What does life need to live on the earth?" The yellow soul tried to analysis the requirements for life.

"Life needs a stomach to fill, which gives strength to the whole body. It needs sex to give birth to new life, and shelter to save it from the changing conditions of nature," the yellow soul said.

"Every life on the earth has almost the same needs to survive on the earth. All creatures have found the resources to fulfill these needs of life. Every life is thankful to nature, as nature's work is not only this. Life is the beauty of nature, but nature is immortal and life is mortal. Nature is not tied to earth or planets only; nature is above that, for which a life is not enough to understand or judge," the yellow soul said.

"Life is praying, life is enjoying and life is facing, but the process of nature is one and the same for all of creation. Nature has enough resources for life. Life cannot blame nature for any scarcity. One can blame the others for capturing or possessing nature by use of power, but cannot blame nature," the yellow soul said.

"The needs of life could be fulfilled easily from nature or with the help of nature, and

97

nature is able and capable to fulfill the needs of life," the yellow soul said.

"Nature has provided enough resources to fulfill the basic needs of life, like sex, sustenance and shelter. All the creatures are satisfied with the blessings of nature in this regard, except man. What man has made of nature and its resources to fulfill the basic needs of life cannot be imagined," the yellow soul said.

"What man has made of sex and how he unnaturally used it, is impossible to believe. He developed his wishes in the most unnatural way. He does not use sex to create new life, but uses it to fulfill his lusty wishes. Man has made his female a prostitute to fulfill his evil and sexual desires. How man compels his female to go to the brothel, no other creature can understand. It's only man who does all this easily, without any fear. No other species on earth do this and use sex like this. Only man can do this unnatural thing, and he does all this proudly and feels no shame about it," the yellow soul said.

"When all the creation depend upon nature to fill their stomachs to nourish their bodies, when all the species over the earth look towards nature to provide food for their livelihood and when all the

98

creation thinks themselves part of nature, the man on the other side developed currency and depends upon money for his livelihood. Without money, man is unable to move on the earth and to take his food for daily needs. Man is a slave to money on the earth. Man all over the earth looks towards money every day and not towards nature, which is capable and provide food for its creatures, including human beings," the yellow soul said.

"Man is not limited to fulfilling his wishes for food or nourishment for his body, but goes beyond to invent so many worldly things that are becoming his catastrophe instead of nourishing the human being. Man has so indulged himself in these things in this way that he has forgotten nature and depends upon these things. He is eager to earn money to get these things. Everything is on sale in the world for human beings, and all human beings are slaves of these things and money. If man has no money, then he has no right to get any benefit of even natural things, as man has captured and taken into his possession everything on earth and has enough laws to prevent others from enjoying nature's bounty. Whoever does not obey his laws and rule, for him man has made ammunition to eat and to die.

"All over the earth, man is daily killing such men, who don't obey the dictated terms. There are wars over the earth between traders of these things, under the name of fake faces of philosophy. Nature is nothing for them, and even on the other hand, man is showing himself the rival of nature and forgets that only nature is capable of nourishing him, not the things developed by him. All things made by man are unable to compare with nature and are not a substitute for nature. These things are enough for man to control man, but not nature. These things are keeping the human being for the time being away from nature and impressing and compelling to live in their control, but this is not final. The man was, has to and is dependent upon the nature—there is no other substitute," the yellow soul said.

"Man has made big buildings over the earth, but is still unable to save himself from the wrath of nature, like quakes, tsunamis, hurricanes, tornados, disasters and storms. He bears and stands with the other creature before this wrath. During the wrath of nature, all man's progress becomes useless; the strength of nature prevails everywhere. Man looks like other species compelled by the circumstances created by the wrath of nature over the earth," the yellow soul said.

"Other creatures do not use the earth as man does to fulfill useless wishes. They like to live in the natural way. They do not have useless wishes like man to possess the whole land as their personal property, as all creatures, including man, are mortal. Other creatures understand that there is nothing immortal, so they behave like that and live all over the earth like that. Man has made the condition of earth so miserable that no one can imagine the consequences of it. Everyone in creation is scared by the acts of man and is suffering the so-called progress of man," the yellow soul said.

"Man has made his food poisonous and contaminated. Crops cannot grow and ripe without poisons and chemicals. Fruit cannot ripe without poison and chemicals. Man is even making the whole atmosphere poisonous and polluted. Man is turning away from nature, and the trader is hiding nature in his manufactured material made by him from the natural resource in which he mixes his own resources to make the material that benefits him, ignoring the health of human beings. Moreover, that material is available only with money, which is not in the hands of the common man," the yellow soul said.

"Man has made the earth poisoned and polluted and otherwise too weak for his trade or to expand his trade. Air and water are poisoned and polluted for his so-called progress. The sky has been poisoned and polluted, and fire has been converted into ammunition for killing human beings," the yellow soul said.

"This is all because man has gathered knowledge. He preserves the knowledge through his coming generations and hands over the knowledge to his coming generations for new researches. Now knowledge has become the major source of his trade and through this trade he loots all human beings, by making the human being weak and addicted to these material things made by him from the natural resources. Human beings are bearing this crisis created by the trader people everywhere on the earth, and the trader is killing human beings to maintain his rule by trade," the yellow soul said.

"The human being is entangled in the net of progress of traders in such a way and forced to follow the law of traders to keep the human being under their control, so that the human being is unable to get itself free from this crisis. Human beings have the right to express themselves only in a peaceful way under the shadow of ammunition, when they have to eat ammunition if they go against the wishes of traders," the yellow soul said.

"The problems of human being are so big he cannot see the blessings of nature. Now if the human being even tries to look towards nature he will not be able to enjoy nature freely, as it is not acceptable to the trader, who is able to keep the human being dancing without facilities, having controlled and possessed all the resources of nature and even the major part of nature. There is not even free land for the human being to use. It is all under the control of power, which is available only with the money developed by the trader. No human being can live freely. They are bound to obey the laws of traders and live under the dictated terms of the traders," the yellow soul said.

"Men's progress has gone beyond the needs of the human being to the destruction of the human being, and everywhere they are fighting to take control of the resources of nature and to prevent the human being from using them use freely. They are determined to sell it under their trade names. For this, they are killing and playing with the blood of the human being under different arrogant slogans to impress the human being and defend their ill will. Nature, creation and the human being are suppressed under their ill will," the yellow soul said.

"Man has developed dangerous weapons to suppress the human being and to bring him into the fold of trade completely and to crush all his wishes of freedom. The human being is bound to submit to the dictated terms of traders and is bound live according to their wishes. The human being is bound to live coarsely under the shadow of the power of trade. There is no way for the human being to come out from the lust of progress, which is not even his need," the yellow soul said.

"On one side, the human being is living without the natural resources of food and shelter and is bound to live without the basic facilities in the dirt of progress, and on the other side, food is being wasted under the control of trade. The trader has changed the shape of natural resources and stamped his trade name on it and is selling them at his price, ignoring the needs of other human beings. The human being is being prevented by traders from using natural resources freely, as they have controlled and possessed all the resources of nature. The majority of human beings are dispossessed from the use of these natural resources," the yellow soul said.

"Man knows his own destiny but my path is towards west from east. The sunrays daily go from east to west with me over the earth. The sunrays daily give new hopes to be busy in their routine of life. Life starts its routine with the rays of the sun," the yellow soul said.

A lion roared, but his roaring was not like the king of animals in the jungle. The yellow soul thought the king of the jungle must be in crisis. The yellow soul knows the crisis of the king of the jungle, but it tried to hear this from the mouth of king of jungle.

"Jungles are in the possession of man, in which other animals are kept by man for the beauty of the jungle, and some lions, just to maintain the conceit that man is so liberal. But the boundaries are limited, in which lions are not kings, but dependant on man's mercy," a fox said.

"Lions were aware of the values of life. The lions forever maintained those values. The lion would not have got the designation of king of the jungle if he had not maintained the values of life. Everybody knows the lion is a carnivore like most men, but the lion also values life. He eats flesh of one animal and hunts them one at a time. After filling his stomach, he doesn't kill or hunt other animals to hoard food or for unnecessary purposes, such as to show his power. When he is sated, no other has fear from his power. But man is dangerous whether he is sated or not. Man is more dangerous if he is sated. Life has more danger from him if it is his pet. He stores the flesh of his pets for his future. He eats less, but destroys more needlessly. Life is more scared by the acts of man than the king of jungle, which has become a show piece in the zoo of man," the hare said.

"Man has forgotten the values of life and is even killing more men than animals through starvation, weapons and ammunition all over the earth. Nobody is safe in his possession over the earth. Even the earth is not safe from his acts. He is trying to find other lands in the sky and discovering more places to control and possess. Nobody on the earth can understand the determination, aims and goals of man.]," the sparrow said and fled away.

The yellow soul wants to hear some from the king of the jungle, but after hearing others, it ignored the roar of the lion and made its way again through the mountains, plains, sea and many other places where life was living in the lap of nature to find its companions.

"Several ants were dying with a pinch of poison, as man does not want their anthill near his dwelling places. Several insects were dying from the spray of poisons, because man thinks that those insects are injurious to his crops. Several birds were being killed, because they are harmful to his crops. Several were becoming victims, to die without any cause." The yellow soul saw, but not taken notice of it, thinking it's the routine of their destiny.

"Once I went in the proboscis of an elephant and gave it a cut in the nose. The elephant was tossing badly and snorting in a pitiful condition. On that day I felt that I am not so small to live on the earth, and I am able to teach a lesson even to an elephant." The yellow soul remembered the time of her life when it was alive.

"In the same way, so many times to kids and sometimes to elders I taught a lesson for their negligence towards its presence." The yellow soul remembered the important and risky moments of its life.

"A beautiful anthill made by the unity of ants, in which it spent life's important moments, in which even rain water cannot enter, was very safe to live in, and harmful to none except man's thinking. The unity was great, if the lead found something, we brought that with the help of each other for us for the mother ant in the anthill." The yellow soul remembered more beautiful moments of life.

"The benefit of a small body is that it can go up the proboscis of an elephant, but can save itself even from under the foot of the elephant." The yellow soul was still remembering its lifetime.

"Millions and billions of sparrows were flying from one place to others places for food. Maybe due to the change of season, this place was no longer secure and the season was no longer suitable for them. The other places where they were shifting might be more secure and suitable for them for food and living purposes. Their communication system, search for food and suitable living places according to the season, information to all, decisions helpful for all and the decision to migrate unanimously are the important aspects of their living standards. In the same way when it was alive like this, ants' decisions had also the same importance." The yellow soul was watching the earth full of life.

"Man's society under the burden of theories, ideologies, atheists, theists, castes, religions, countries, races and so many other factors is being ruined and is becoming a crisis for human beings. Nobody is ready to fight for the human being. Everybody's thinking has become that of the slave. Nobody is ready to think for the benefit of all human beings. The freedom of sparrows, decisions of sparrows, work for the benefit of all sparrows and no war of any type for rights and duties again have beaten man's sociality and society, the yellow soul observed.

"The whole earth was looking beautiful, full of life and nature, in which only the smoke of progress was not impressive to all of creation and was creating discomfort to all." The yellow soul said.

"If the butterflies, flowers, birds, animals and trees are making the day beautiful with the rest of creation, then far from the dazzling lights of the glow worms and fireflies, the twinkling stars and moonlight are also making the night more beautiful and calm. Creation and nature were engrossed in the daily routine. The music in nature of air, water, trees, rivers, waterfalls, streams and sea and bays are the other beautiful colors of life and nature," the yellow saw.

"Somewhere, the rain bird was praying for rain, and somewhere the rain was pleasing the creature. Nature was looking at all this with life distributing its blessings everywhere," the yellow soul saw.

"Sunrays were making the water of the oceans hot, converting it into steam. And then the air started to play with it, sending the clouds hither and thither in the sky, which welcomes all open-heartedly, and returned it to the earth again to please the all creation, which needs all this to nourish life. The water in this process somewhere becomes the snow over the mountains and other parts of earth, somewhere looks foggy, somewhere becomes water to flow in rivers and streams on the earth, and again, after fulfilling the wishes of life, all over the earth falls into the sea. How nature is playing its role free of cost to nourish life," the yellow soul saw.

The yellow soul was still proceeding towards its mission, but suddenly it felt that it was not possible to find the other souls now. It felt some changes in behavior. What was happening or what was likely to happen, was not apparent to the yellow soul. It stopped doing anything and became calm.

_8_

The blue soul found itself in the darkness. Many days passed in the darkness. It was feeling the change. It was feeling that it was transforming into something. It was feeling surprising new movements.

After deep darkness it felt that it was in the body of someone who was trying to come out from the darkness. It was growing into the body of someone who was getting life. It felt that its body was still incomplete, but it was growing and developing into a life. It was feeling that there was light outside the walls of darkness. It was feeling that there was big movement outside the walls of darkness.

A day came when the walls of darkness broke, and it saw that it was not only a soul now. Its identity as a blue soul was finished and a new identity had been formed.

Birth is not only pleasant movements of life. It's painful, too. Liberation from the walls of darkness may give much relief, but still it was not free from all the difficulties of life. Experience of life is very difficult to get. The birth of someone from the darkness to the light is a very important movement of life, which it was experiencing.

The journey from the soul to life was a journey from light through the darkness towards the sorrows and happiness of life.

It got the body to live the life. What it will be, or what it will give to it, was contained in its knowledge, but for the soul to get a body is life, and now it has to live in this body and bear the fate of its body.

In whatever body a soul takes birth, it becomes the identity of that life and bears all that which nature provides to live in the creature of other lives. Now it has taken birth and its destiny has been started.

What it will be was in the hands of time, but now, when it took birth from the egg, the upper part of egg was broken and coming out was the first step towards the new life.

I was within the circle of the female snake's body. She brooded up me to hatch but left me at my own fate without caring for me. In what direction I started the journey of life and what to do in the future to keep myself alive were up to me to decide.

I started to run to save my life, without legs, moveable eyelids or external ears. My teeth were not to chew or cut anything. My first run was to feed myself, which is the first and last necessity of life and without which life is meaningless. Who will teach me all this to start life? No one, not even my mother or father is there to tell me anything. I have to start my journey by my own efforts.

Before starting the journey of life, I was well aware of my previous birth and the journey of my soul. Perhaps I will forget all this in the pain of the journey of life, but now I am afraid to remember my previous birth. I fear that the shadow of my previous birth will affect my present birth. Why I am feeling so scared, I do not know, but I cannot forget the reason.

When in the previous birth I was a dog, I killed a snake out of fear that the snake might harm a child of my owner at that time. There was no regret at that time in my mind and I felt proud of that act of securing the life of the child of my owner and of killing the snake. But now, when I am in birth as a snake, I feel scared that the snake's soul must have taken birth as a dog, and the dog will kill me in the same way. It may be that that soul has become a man and that the man will take the same revenge of its previous birth, or it may be that the soul has become a hawk that will kill me mercilessly or maybe a mongoose, my permanent enemy.

Maybe it is only the fear of the shadow of my previous birth, and nothing like that will happen in my present life. It is just the start of my life; maybe the shadows of my previous birth are affecting my present birth, which never happens in reality.

When I had eaten the first insect to feed the life I became torpid, while the process of digestion was taking place in me. It was my first prey to keep myself alive, which actually is the necessity of life.

When I awakened I forgot many things from the previous birth. My conscious was not under any fear of the past birth now. The journey of life is so difficult that one cannot remember the past birth. The present becomes heavier than the past.

What are the virtues and vices of life I do not know, but my first food helped me to forget the shadow of the previous birth, and maybe it will help me to live in the present without fear of future. Virtues and vices are laid down in the mind of one who does not know how to live in the present. The present introduces you to reality and the values of life above the past and future. I do not know what will happen to me in the future, but my present helped me to forget the past, as at this stage my first duty is to save myself. I do not know if I can save myself or not, but I am trying to do this above the shadows of the past and not worrying about the darkness of future. Now what I feel is that I have love for my life and I want to save it.

It was my first food of life. Neither my mother nor my father was with me to tell anything about how, what, which or where to get food, good or bad, vegetable or non-vegetable, virtue or vice, healthy or non-healthy. I was all alone to survive, and it was not easy to survive.

Now nobody was with me to tell me how to start the journey of life. My first food freed me from the worries of the virtue and vice of my previous birth, as well as of my present birth. I threw away, too, the burden of the shadows of my previous birth. I have no time to think anything of my previous birth, as my present has become too heavy on me.

I rose above the contemplation of what is sinful or what is virtuous with my first meal of life.

Now I have to struggle for food beyond these thoughts to survive. My first meal became the scale to measure the depth of my destiny. Now it was out of my hands to judge the fruit of my sins or virtues. My first preference was to survive, and for that, there was nothing important before me in considering the values of virtues or vices. I needed my food and there was nobody to provide me anything, neither my father nor my mother. It was for me to decide how to survive. That insect was before me, which I have eaten as my first food, without thinking at all whether it is sinful or virtuous.

Maybe my first food settled many things in my body for the future. Maybe the function of my body gave me the desire to eat such food, as the digestive system in my body could not digest most plant matter. That's why I am carnivorous by birth. Nature's other gift to me was that one feeding was enough for several weeks.

Time never stops. It is life, which becomes by time birth to young, young to old and old to death.

It's also the first time when I was growing, and I have shed the outer layer of my skin. A new layer developed below the surface of the old one. It separated gradually in preparation for being shed. I began the shedding process by rubbing my nose against rocks to separate the old layer from my lips. When the old layer loosened, I crawled out from old skin

It was looking all good.

-9-

It was darkness all around. I was a red soul. Now it was looking as if all that was over and I was awaiting the light of my destiny, lying in the dark. I was being brooded in the incubator away from my mother. It would take twenty-one days. May be I was not being brooded by my mother, but I was in a well-maintained temperature and I was being well hatched up. I was in the egg of a hen being hatched up in the incubator by the poultry farm owner.

After twenty-one days when I saw the light, perhaps I could not see my mother, but I was in good hands, which were taking care of me. I think my life was not so important to me only, but it was important also to that person who was taking care of me. I heard the peeping of other young chickens like me, who were being hatched up with me.

Very good care and the darkness of twenty-one days never allowed me to remember any past moments. There was nothing in my mind of the past. I had been mixed with other young chickens since birth, and I was enjoying life with them.

I was fully dependent, but regularly getting good food like the other chickens. We were being treated as important guests. Everything, including every treatment, was being done in good time. Medical check-ups, cleaning, feeding and all our other needs were being taken care in a timely way.

Water and feed were always available for us, and we could decide how much we needed and how much we can ate. It was available for us without any efforts by us. It was our owner's responsibility to worry about everything to do with us.

All around there were chickens and chickens, and all were enjoying as we were, and we all were happy looking at each other.

We were in a cage and had to stay in the specific area. Nearby were the houses of people who had some hens. Those hens were free and could go hither and thither at their own will, but had to find their food by their own efforts. Their owners only provided them with a little bit of food. Their freedom sometimes hurt us, but the free availability of sufficient food didn't allow us to feel much pain from their freedom.

They were healthier than we were. That was due to their daily exercise, as they had to go hither and thither to find their food.

One day we saw some children who were running after a cock. Cock was running very fast ahead of the children, from one side to the other, to save himself from their clutches. They caught the cock. I was not aware of the reason they were catching the cock. Some elder chickens told us that their owner must have sold him to some other person, who would cut the cock up for meat to add to the food.

I tried to show sympathy with him, but was surprised to see the elder chickens were smiling at me. I didn't know the meaning of the elders' smile. When I could not understand anything, one elder chicken told me that we are made for it. Our owner is not giving us the delicious food for no reason, and it was not an accident that we were in the cage. When he sees we are ready to go, we have go to be cut up for the food of human beings. They like to eat our meat in their food.

I could not sleep that night and was restless whole the night. The next day I saw whole cages of elder chickens being vacated, and some persons were putting the elder chickens in some empty boxes specially made with small holes. They were aware of what was happening, but looking helpless, not protesting too much and were taking their places to sit in the box calmly.

After two or three days, some persons put more young chickens in their cages. They were peeping, taking new places to live, worried less for their future, as they were being kept from the very beginning very carefully in a well-maintained temperature, with good, healthy food specially made for their health and digestive system. They were looking very happy, excited and satisfied with the well-behaved and well-maintained atmosphere around them, even without their mothers.

I saw outside some birds flying freely from one tree to the other and from one branch to the other. I took the measure of my own wings. These were like those of other birds, to fly in the sky. I had already seen outside hens and cocks that also could not fly like birds. The cock could not save himself on the strength of his wings. I remembered the words of my elders that our destiny is to fulfill the food needs of others. Perhaps nature has made us for that. To fly in the sky like birds is not our destiny.

I do not know why I am feeling the life outside my cage is good. I do not know why I am thinking the life of hens and cocks outside is better than ours, in spite of the fact that the destiny is the same of all of us.

The freedom of other birds is not for us. But I do not know why I am searching for the meaning of freedom. I do not know why I am not conceding the destiny of my life. The fear of death may be the cause, but love of life beyond expectation is also not the fun of life. You have to face the reality. I do not know why the wire of the cage is looking to me like the enemy of my happiness in life.

Somebody outside has come to clean our dropping waste. All humans do not look the same. The expressions on all the faces are not the same. The skin of all the faces is not soft and smooth, some faces looks sulky, with dry, poor skin. The difference is written across the forehead of everyone, which is readable easily without any hesitation. We can imagine who is very fond of our breast and legs and who satisfies himself with claws and veins.

The man with the nose mask who was cleaning the waste did not look like the same personality as the man who owns us.

The peeping of the new young chickens was pleasant, as they also looked happy with their warm welcome. The growth of our bodies from eating the healthy feed looks very good, but does not explain why I feel fear today from my own growth. Still, I have to survive with or without the fear of death, until my last breath, as I love my life.

I am too young and I should not take too much fear in my mind. Facts cannot be changed at any cost. We have to pay the cost of life with great taste.

Another man has come with fresh feed. We all were waiting for that. It's our time to feed.

-10-

The first meaning green soul took from the hybrid was that it is my destiny, due to sins of the past birth. I took the expression from the men standing around and doing all this, which in my view they were not doing due to any enmity. From this crossbreeding, their wish was very clear: to create a powerful animal that would not have a sharp brain to think, would but have strong muscles to carry heavy loads.

I have to pass through the semen of a donkey into the womb of a mare. The mare shall give birth to the third race, which will not be a donkey or mare, but a mule. The secret of the mule's power is that she remains sterile her whole life, except in rare cases. The reason for the sterility of the mule is that a horse has sixty-four chromosomes and a donkey has sixty-two chromosomes; so, the mule is left with sixty-three, an uneven number that does not divide into chromosome pairs, making the mule unable to reproduce.

If in a rare case the mule gives birth to a foal, then that foal retains extraordinary power.

Some says it's a myth, but some examples show it to be true. There are only a few examples that can be counted on fingers in the whole past history. In a real sense, the truth was that I had to remain sterile for my whole life, and that was my destiny of life.

It was my own false notion that there was any shadow of the past birth in the process of my present birth. It may be their need that was inspiring them to do so. It looked as if it was not due to any enmity of past birth. Nobody knows at the time of crossbreeding which soul will accept this body that is being produced by this crossbreeding. Sometimes your past shadowed your present with false notions. There was nothing like that in their power. It was not in their hands to influence it. It was the rotation of nature and the life circle.

The persons standing around watching, who were laughing at their achievement, left that place after the satisfaction of my father, whose lust actually was the root of my birth. I passed through the lust of my father into the womb of my mother.

My mother gave me birth as a mule. I was not a mare, horse, donkey or jenny. I was a mule, a different race from all these. I have to live with this different identity now. However, there was no difference in the love of my mother. To my mother, there was no difference whether, as a foal, I was mule, horse or mare. Her love for me was sacred.

My owner was very happy with my birth. I do not know in which eye he was looking, but he was happy. I think he was happy by observing my price, the price of life, and that was the only reason for his happiness.

Who knows life's destiny? The destiny of birds and animals may be somewhat different, with less effect on the future by those that are away from the grip of man; but destiny in the hands of man has much effect on the future. You may say it is the progress of man. One has to believe in destiny in the hands of man's progress on the shoulders of his knowledge. This development of knowledge by man has made destiny terrific, not only of the mule or other birds and animals that are living in the hands of man, but including man, who is also a fiber in the dilemma of the entire tapestry of a system created by their bosses, the owners of their destiny.

In the system of man, one has to stay on course with the destiny of their life. So I was not different from the play of destiny. I was also not aware of my destiny, of what will happen in the next moment.

Anything can happen, good or bad, as all of it was the play of destiny now, because I was entirely in the hands of my owner. My destiny also depends upon my owner's decision. His one refusal or acceptance can change the destiny of my life. Whether I have to carry the load of man from one place to the other for the whole life for which he produced my sterile life, or whether there is any other haven for me, like the dreams of a poor person.

Belief in destiny, I think, is born from the relief one gets from the curse. Any relief obtained in life from the hell of life makes firm the belief in destiny, because it looks like a miracle, especially to the weaker section, which is unable to presume anything good to happen. If I do not believe in destiny, then it may be my foolishness, as my birth was made on the presumption to carry the load of man, and if I do not do that for my whole life, then how can you deny the play of destiny?

Knowledge, if it creates troubles for life for some, on the other hand gives pleasures to others, too. Pleasure may be only the mockery of one who has been compelled to stay on the stage of the world to play his part of life, but one has to dance on the fingertips of one who is the master of your life on earth. Your master may be under the possession of another powerful master, but at present, you are tied to the wish of your present master. Its knowledge that makes life a drama and one has to play the drama of life with the stick of the system, a safe haven for some and hell for the rest.

The birth of knowledge came about because of needs and the needs gave birth to greed. Greed became trade. Trade prevails over the world and makes life a mockery. Life is still there waiting her turn in the queue of the curse of knowledge.

Knowledge gives shape to life in conditions that ultimately become the wishes of life to live. False wishes gave birth to slavery controlled by greed. Greed gives birth to power, which kills wishes and sets life in a race to achieve false wishes.

Knowledge of man has given dynamic shape to the world hurtling towards destruction, with unnecessary rapidity, destroying the sources of nature, by making nature and life a victim.

Control of my life was entirely in the hands of man, specifically my owner. Yes, it was a time of change of ownership. A company was buying me to send me to the zoo of a foreign country, where I will be a distinguished thing to see. In this way, my destiny saved me from carrying loads, maybe at the expense of the loss of some other freedom.

-11-

Life starts from dark. After dark, you get the light of life. Dark and light, both are part of life. Life starts from dark and ends in the light.

My darkness has been started. I, the yellow soul, am entering into life again. I am landing. My landing place is my mother's womb, where I will learn the meaning of love. What is love? Love is the womb of mother. No love is greater than this. The relation of love from soul to soul starts from here. No love surpasses this.

I have seen all aspects of life in the womb of my mother. It was a good place for me to learn about life. Actually, it was not our field to discuss. I think birth is the result of our thinking about life. Life is to live, not to think about. We cannot think about life. Thinking about life is useless. I now realize this, that our views about life were useless. We were ignorant about the reality of life. To face life is important. We actually were unable to discuss life. Life is not to discuss. Life is to live.

In the womb of my mother I have learned only this thing. Now I feel shy about our useless discussion about life. The reality of life is above discussion. The warmth of love I am getting in the womb of my mother I could not explain anywhere in any discussion. This is not even a matter to discuss or explain in words. This cannot be discussed or explained in any way. The reality of life cannot be explained.

There is lot of difference between expressing views about life and facing the reality of life. If anybody sees the reality of life, it does not mean he understands the same up to the depth of one who lives life.

It is true that life is the creation of nature. But to live life in nature becomes a more personal experience, which differs from overall aspects of life. Now I am feeling a more personal experience. I am facing the reality of life from very close up.

After entering into the womb of my mother, my viewpoint about life has been changed. My feelings are being affected by the feelings of my mother. Life is taking birth from life. Life is becoming the source of life. Life is nourishing life.

In the womb I am taking the shape of a complete body. Every part of the body is being developed and starting to function, which is my life.

It is the start of my life. It is the beginning time of life.

I grew up with the warmth of love of my mother for nine month and then I took birth on the earth. With the first crying I started the journey of life on earth. My first crying was the sign of my good health. My first crying was due to the natural introduction to air, which entered into my body. Now this breath gave me independence from the body of my mother. Prior to this I was the part of my mother and from nourishment to breath I had to share with my mother. Now I was independently enjoying nature, and my body was able to accept the blessings of nature.

With the first breath, with the first tears and with entrance of air the first thing that happened was that I forgot many thing of my past, and I was introduced to new things of the world. The love around me was also taking away from the effect of the past.

It was the start of polishing my brain with the worldly things on the earth. Every stamp on my mind was the experience of my life on the earth. Every stamp on my mind was not only helping me to forget the past, but also channeling me from nature into personal doings and difficulties. I was indulging in worldly things, the love of surrounding society filled by relations and their circumstances.

Now I can recollect very little of my past, when we decided as a soul to discuss life, and we were separated by the light and started to find the reality on our own. I do not know when I was put into the womb of my mother for nine months. I cannot see my other companions of the past. I do not know where they are. I do not know in what shape they are. I do not know why we started to discuss that. Now I am facing reality. The reality is more difficult to bear. I am being limited by the life. The love around me is taking me in its grip. I am developing into the circumstances of my surrounding. I have to fight for my identity and status in and around the world of mankind. The struggle of life will finish forever all my memories of the past birth and as a soul, and I have to start living in the present circumstances entirely.

Around me there are circumstances of my family, of my neighborhood, of my relations, of my state, of my country, of my culture and of my society, in which I have to prove myself in the competition of life, and which will cover my life as a whole, and I will forget everything that I was in the past and what I will be in the future.

I have to face the stored knowledge of human beings, which will realize my identity in the world. I have to seek my place in the progress of man, instead of the blessings of nature possessed and captured by some of the human beings.

The face, with the looks of childhood, will be polished with traditions, rites, languages, laws and rules, progress of knowledge, ideologies and circumstances. I have to bear the burden on my shoulders like others around me of this.

I will be prepared and educated to face those realities, which are facing others around me. They will introduce me to all the difficulties and ways of life. They will tell me how to live in the progress of knowledge and what I have to face in the power of knowledge, and what I have to acknowledge to live in this society.

My polished face will become like others who are pulling the load of life in the powerful world of knowledge, captured and possessed by powerful holders of knowledge tied in the strong system.

I will not be able to be thankful to nature, which nourishes the whole of creation, and I will be compelled to be thankful of the owners of the world, who have captured and possessed all the resources of nature.

I will have to stay away from nature, in spite of the fact of living in and on the blessings of nature.

I will have to accept the light of progress, instead of the light of nature; to accept the earth in shoes of progress, in spite of the reality that I was born and will part of earth after this life; to drink the water of progress, instead of to be thankful for nature, which nourishes the whole of creation with the rotation of it; to hear the polluted noise in the sky of progress, instead of to take breath calmly in the sky. I will have to borrow the breath from progress, instead of flying with the air to enjoy nature.

I will be polished with progress and shall be made to hound to get the benefits of progress. I will not be able to remember nature. The power of progress, which captured and possessed the resources of nature, will be paid for to me with the health of my life.

Love around me will tell me all the ways in which they have learned from the experience of their life, how to run and accept this atmosphere with struggle, which I have to adopt. My brain will be stamped and confined in this struggle of life.

Days have passed and the bag of books on my shoulder is the knowledge to wash my brain and to keep away from nature, and learn the progress in which I have to live and pass the exam of life.

-12-

"It was our foolishness to discuss life. We were separated by the light, not only because we were doing something wrong, but as a little warning to us to forget the past. But we still were adamant about discussing and finding the ways of life, in spite of the truth that these were our false notions, depressed and shadowed by the burden of our past in the previous life" the multicolor soul said.

"All my efforts to search for my other companions are gone to waste. Now they all have taken birth on the earth, but I am still a soul."

I can see all of them, but they are unable to recognize me. In life they are unable to see me. I have no way to make myself known to them. I am a soul and they are in bodies of life. I have no body. They are not able to recognize even each other. They are now fully living their lives and have no knowledge of the past. That's what they are talking about; their pasts are the shadows on their undeveloped brains, and they will forget everything with the introduction to the life's routine.

I can see the snake searching for food. He will take the life of someone and will make his victim. He will sleep for a long time with satisfaction after feeding, and will try to save himself from being the victim of another life.

I can see the chicken in the cage, very busy in eating the food and getting ready to be the dish for another life.

I can see the mule, another life understanding the meaning of life in the hands of life.

I can see the child with his bag of knowledge, getting ready to give shape to his life and trying to adjust to the world of knowledge gathered by the human being.

Apart from this, I can see the whole world. This is the real rest of life, which I am enjoying.

-13-

I was digress for many years and was thinking this was the end of my transmigration. Many lives were treating me as a bad spirit and were frightened of me. I was not very scary, as I had no thought of revenge against any life coming from its last birth. I had never even tried to frighten any life. I don't know why I could not get any birth like my other soul mates. There is only one soul whom I can perceive at this time. The other souls have left the bodies again, and I cannot recognize them now.

He is in the ship going to the other country with some other companions. Atrocity and fear compelled him to leave his country. Sitting in the corner, he was thinking.

"Man is not satisfied with committing atrocity on the trees. Man is unfulfilled with the atrocity against birds and animals. The lion does not make the birds and animals pets for his food. Maybe nature is not in the control of nature, but what more can it do to fulfill the wishes of man?"

"I think man is not ready to admit that he is indigent before nature. From life to death nature plays its role to educate the human being, but man's pride still races towards destruction. Even if man's activities cause the earth to disappear, then what will be the effect on nature? The stars, sun and moon will not cease to put their light in the sky."

"My mind is flying. The birds are flying. It looks as if we are enjoying together the beauty of nature. I am trying to join their flying moments. My mind is also flying with them."

"Whoever uses power to change the ways of life is always remembered as a tyrant, and whoever uses peace to change the ways is remembered as incarnation in history which is enough to witness."

"I was so busy in my life's routine that I could not ask flowers what happened to them. I only watched their destiny. From life to death they sacrificed their lives for man. I don't think about whether anybody remembered that or not. "

"My mind always liked to fly like the birds, but failed to think how many storms come into their lives."

"I was the fan of the beauty of mountains, but could not see the holes, burrows and volcanoes of mountains. I never imagined how strong they were to serve the universe and add their share to natural course of things. I also don't know how much marble and costly stones, coal, iron and other expensive things, which were their power to run the universe, man have taken out to fulfill his selfish needs."

"I always tried to see the depth of the sea, but forgot its tsunamis, hurricanes and storms. I don't know how much of its strength man's selfishness has stolen and how much he has poisoned it."

"It was good that I never examined the depth of the sky, as I don't know how many suns, planets, earths, and seas are wandering in it."

I told a bird, which was looking at me, that man isn't afraid of hot seasons, because he has invented air conditioning for his home, car and office. He doesn't care about cold seasons, as he invented everything to save himself from every disaster. It may be fear in your wings to save yourself from the hot or cold seasons.

The bird laughed at me and said that you, the man don't believe nature, so you have become sick, and your invention is the cure for your sickness. I depend solely upon nature, and I can't tell you how many ways nature has kept open for my happiness. I have seen man's destitution in one tsunami when his condition was worse than mine, in spite of his inventions and discoveries, which we have not.

Life gives ways to life. Somebody said break the relationships; there is selfishness in them. Somebody said break the family; it gives birth to selfishness. Somebody said break the society and make a powerful system that is enough to control everyone. They forget that the way to be human is through relationships, family and society. To kill the evil, you can't kill yourself, your relationships, family or society. Kill the evil only.

There are lots of traders, but there are very few who trade their lives for others. Everybody wanted to take that path, but it is not easy to trade life for others. Whoever traded life for others, the real traders really hanged them, because those traders' lives educate the people to save themselves from the clutches of real traders, who trade their blood for the things they want. The bevies of birds, the flood of human beings all around over the earth and sky are looking good. In the home and the nest, there is almost the same type of struggle for life. The fairs of human beings and the flights of the bevies of birds look very common. Life, struggle, then go to die. Maybe I am thinking to be a bird more than a social animal. Man also has taken so many flights of progress that's why I am also thinking like that. But when I brag about my power, I compare myself with the lion, an animal. Now keeping in view his dubious power before man's progress, I am afraid to use him for comparison, any more. Nature is laughing at my wavering mind.

Will the vanity of the progress of man ever be able to reply, that's what nature has done, that's what nature is doing and that's what nature will do for life and for human beings. The whole of progress may vanish in a moment in the process of nature. Man and his progress have no importance before nature. The extremism of man's progress has no meaning in the eyes of nature; it's the mockery of man's before nature.

I would have told you the meaning of life, if I were able to do that. It's not in my hands. Sometimes cross fixes the meaning of life. The harassed people, whom you see busy in their small happiness, die so many times in a day. I nourish in them.

I asked the bird to tell me something about the seasons. He told me that he is so busy enjoying life that he has no time to think about life. He told me that he will face every moment, whatever comes before him. He said that he knows the capacity of his body, and he is able to fight with every difficulty till the last moment.

Then I asked an animal to tell me something about the seasons. He said that you are also a social animal. What more can I know than you?

I gazed towards the sky. The sky said please don't ask me any questions, as I am not alone in making the seasons. I have so many earths, suns, stars, and moons having different situations, and your earth has its own control over all the changes, according to the time.

I looked towards the earth. The earth said that man is trying to estimate my age now, so what else can I tell you?

While I was thinking how to enjoy life, they came, enjoyed life and were gone.

Life is not so cheap that it comes and goes routinely. Sometimes, it leaves its impression on time.

Time doesn't care. It's life that stamps it.

Selfishness controls the universe. Man, animal, bird, insect or reptile—all are the food of each other's selfishness. Be brave, struggle and save yourself.

Man's discoveries show that big bodies of life are the hybrid of small reptiles. When the earth came into existence, all were small reptiles and later on with healthy food and time they got heavy bodies. Now it looks as if man's knowledge and progress are trying to make them reptiles again.

Man is becoming the house of diseases in the hands of the progress of man, which is not possible in the hands of nature.

While I was finding the meaning of life from the views of theism and atheism, the other creatures were enjoying life.

After wearing life's spring, life's rain, life's autumn, life's winter, life's dry and moist air, life's sultriness, life's puff, life's warmth, life's seasons, life's days, life's nights, life's dark, life's light, life's thirst, life's desert, life's mountains, life's fertile life's wet, life's difficulties, life's sorrows, life's happiness, I could not understand what is life and what is the reality of life, but death told me the meaning of life.

It was my soul mate thinking in the corner of the ship. Now he is singing in very low tone:

That soul has no rest who has a restless soma

Who has chosen the difficult path, and selfless targets?

Another guy who was hearing the words of my soul mate's song also started singing:

Who gets coffin under armpit all the time

Who defeats the atrocity, and is tireless

A second guy who was also nearby joined them with same tune:

Every time a flower has to sacrifice his life

May be a time of honeymoon or time of death

Another guy also sang the song in the same tune:

In whose heart compassion lives and knows to die

Who wrote his story with blood and doesn't regret

All smiled, looking each other.

"Hallo! Brown," the other guy said.

146

"Hey! Black," my soul mate said.

"I am White," the second guy introduced himself.

"I am Crimson," the third guy raised his hand.

"For whom are we singing? Everybody is divided here. Here are Eastern, Western, Northern, or Southern, theist, atheist, black, brown, white, blue, this country, that country, this state, that state, this race, that race, rich, poor, middle class, this religion, that religion. Color of skin and faith has become important, but the color of blood and man's basic needs nobody will try to see." Black was expressing himself until the bird's interruption, "Don't mind, the color of my blood is also red. Add all creation to your thinking, as all creation shares its rights in nature."

"Oh! Good. But man has adventured the groups of the blood, and traders will divide the strength of human beings on the basis of those groups of blood.

Don't think they will admit their defeat if we will join the whole creation in this struggle. They have already defeated the whole creation, which they kept alive only because they cannot live without that," Black said.

"I mean, don't question only humankind; all creation and nature are crying. Those who believe in the law of land are in a mood to extinguish the land, and those who believe in nature are putting nature in question," the bird said.

"I know. But you know the meaning of freedom. You know the meaning of nature. You know the meaning of land. Over the land in the kingdom of nature humankind forgets the meaning of everything. Cruelty left on their mind only the name of cruelty's producer. So they are not in a condition to understand anything of what you are saying and want to tell humankind and me. I was talking to the sleeping and unconscious human beings, dying under the burden of things that are made with the flesh of nature. Don't kill your mother, I was crying before the human race. I was asking them to wake up. There was nothing in my views partial from nature and creature," Black said.

"I have stories like yours. What do you want me to explain? A bee told me the history of honey, which she heard from family members. It was a hereditary story in her family. Their families collect honey from flowers for their future needs. But man found their secret. The man's cruelty was not that he discovered the secret of honey and dispossessed all the bees of drones for his need for honey all the time, but the dishonesty that he is cheating his men. They invented this supernatural thing for their generations and advertised their success. But they were unable to maintain the demand now. Then their search started. They fed the sugar that they had already invented to bees, instead of flowers, which take time for that service, and in quick service, supplied that honey made from sugar through the bodies of bees. But the bodies of bees were nothing. Only their feed was sugar. Bees were happy to get the easy food. But men were dying with sugar, those who had already been given too much sugar for their weaknesses. This was not the bee's tragedy. The bee's tragedy was that they were surprised that there were more bottles of honey in the stores than their population could make," the bird said.

"You don't put me in question, because my observation was only for human beings, and I have anywhere ignored the whole nature and creation. I, the whole creation, am nothing before nature. Only nature will exist. If humankind did wrong then there was the rest of humankind who stood against them."

The bird interrupted and said, "You again limited yourself. I will again tell you the other story of the cow, which she told me. She did not understand, herself, whether her body was machine or body. From birth to pregnancy, from milk to death, up to beef meat she has worked through machines and is unable to see her infants and lover. She is also the part of your saying," the bird said.

"I still believe what I am saying. You are a bird and you can but fight for anything. My knowledge is not limited to human beings; it will work for the whole of nature. However, I have respect for your views, knowledge and pain for all creation," Black said, and became silent upon hearing the voice of song.

"On every step my faith was broken

How I can become the messiah of the world

I am dying by drowning, my friends

How I can become the rain bird

She has put the eggs in the nest of crow

Now renounced by the cuckoo's heart

He fired written death on the bullet

Fire came towards him written death

It's the field of death, death is dancing

What will do mother's prayer if you heard

Air becomes the storm in my courtyard

Will nature save my confidence in the world?

They are in a hurry to win the sky now

Where will he go who lives on the earth?

My request to the crow to return my faith

I will give sweetmeat food with butter"

It is White. The birds were happy hearing the song of White and were tumbling in the sky. He was still in the same mood:

All tells the way to mosque or church

No one tells the way that goes to the heart

Earth losing, sunshine losing,

Moonlight losing, sea losing, sky losing

Man says he is the superpower of the world

Knows everything and he is alert

Humanity suffering, creation suffering

Rivers are running full of blood

They won all sources of nature

Human beings are crying, you can hear

Man of God was standing near him

Requesting relief from sufferings

He was busy, had no time to hear

Made statue of God, became worshipper

The ship was going on and their conversation was also endless.

-14-

I have become so weak that I cannot stay at one place for too long. I could not hear my friends for a long time after that. I don't know what happened to them, where they had gone, where and how they lived and spent their rest of the lives. Then, once again I saw one of my friends.

He was saluting the sea by sitting on its shore, sharing his sentiment: I don't come to your shore only to see how you embrace the earth and to put that love of moments in my heart. There are so many other ways for me to enjoy those moments when you embrace the earth. When the sunshine converts your water into vapor, the sky and wind turn it into rain to embrace the earth. At that time I can put that love in my heart, too. The globe of air, water and earth revolves and rotates around the sun and itself, to nourish the creation and to keep warmth with sunshine. We, the creatures, always see these wonders of nature. How they are nourishing the creatures and the creatures always enjoy life in the lap of nature. There is no substitute for nature.

Men, having or gained power and popularity, took birth on the earth. They proved their power of knowledge, unity and many other traits. What the knowledge wrought today, you can see, but everything is mortal on the earth and water. Everything was mortal and will be mortal. Only, through wielding his power all the time, man showed his wrathful behavior. Opposition to this was always suppressed by power. The knowledge showed its power. Less powerful human beings always were defeated by the power of knowledge. Always, the winner of the knowledge became the hero. Always, humankind was suppressed, and this happened in every age. Nature endured all this calmly through all time.

I came to your shore not only to see this lovely embrace from which creatures takes birth and get the meaning of life, but possibly I have come to immerse the ashes of dead humanity, or I came to vent my pent-up feelings. I am worthless before your infinity. I am a mortal mote before nature, earth, water, sun and sky and their work, whose mind has pent-up feelings, and is otherwise unable to sing and enjoy the vast nature.

On the basis of gathered human knowledge, man has collected and built up heavens on the earth, which he used for his happiness and well-being. With his extreme progress with techniques he has achieved many advances in life, and with this gathered knowledge and some other discoveries he believes himself and placing himself in a position equal to nature. He tries to put the gathered human knowledge on a par with nature and tries to belittle nature, thinking himself the boss who can run nature.

Wearing a mask of the false face of humankind's knowledge, whenever I come near to nature or sit on the shore of the sea, I feel myself defeated. I am unable to enjoy nature as a fan of nature. I am arrested by the human knowledge, dependent every moment on it, have become a slave to it, and am unable to face nature.

When I come here I feel that knowledge is dependent on nature and that nature does not need the help of man's knowledge. Here I feel myself divided. One part of my body becomes natural and the other material. Man collected knowledge from centuries to centuries and developed it to the extreme, producing the latest technology. He made it a wish to achieve, and man became a slave of man, leaving nature behind.

Trade captured nature and the natural resources. Everything produced and given by nature went into the hands of traders. Traders wrote their own name on everything given by the nature, and in this way they abolished the name of nature and became god. Life sometimes comes out from their entanglement on the shore of nature, but becomes confused and incapable. I am not divided, but what I can do about that dependency which envelops me from all angles, so that I forget to enjoy nature in true manner.

Nature has no bombs to throw, but when nature takes a deep breath, that becomes a tsunami. Nature never invades, except to clear the cruelty of man, and that becomes a hurricane. Nature doesn't separate its gases, metals, minerals or oils to waste, but to maintain its balance, when it takes a turn it becomes an earthquake. So, I don't mind if man wants to use his power to confront nature, because only this way can he see whether he is powerful or nature is powerful. Ruining nature is man way of waging war against nature. Who wins? Only time will tell.

I can't be divided. I may be dependent upon the material things of man, but I can't forget nature. Man from eternity was dependant on nature. Maybe to capture the resources of nature, man bombed and killed men and enslaved men like the resources of nature, but nature never forgets to fulfill his duties to nourish creation. It is fulfilling; it was and it will fulfill.

Accumulation of knowledge becomes a step towards destruction, too. Those whom are competing with nature and giving prizes for achievements one day will become part of this earth and nobody will remember them. Their money, gathered and captured resources, will not help them to remain on the earth, but nature will still serve and nourish its creation. One day, the collected knowledge of man shall also be buried in the earth. All that has been captured and won shall not be in his name.

Men know that their knowledge is incomplete. They know they can't go to any other planet to live, and only this planet on which they are living is welcoming to life. They know that their knowledge will be defeated on this planet by the very normal ways of nature, but still they seem determined to triumph over nature.

Whenever nature takes a deep breath, the material that man has taken out for his use from the earth, and by which he has weakened the earth, the earth will take back to maintain its energy to nourish all the creatures, as nature is not the slave of man. The earth is not the personal property of man. The earth can't tolerate the possession of anyone. Everything will be scattered. Everything will be topsy-turvy on earth and water. Where there were earth and mountains, there will be water and sea, and where there was water and sea, there will be earth. Who died and who remained alive shall be countless. Knowledge shall be dumped in a deep grave. This will happen because air, sun, sky, earth and water are not so weak that they are unable to give answer to man's meanness.

Man's blindness towards nature is not new. But now, man has crossed all extreme boundaries. Man shows his sympathy with nature, but blind man doesn't know that he needs the sympathy of nature. Man, who captures the resources of nature by throwing bombs and killing human beings, one day will be indigent before nature.

Sitting on your shore, I am getting some relief, because I am also a slave of the miracle of man's progress. The miracles of man proved to be a trade and business for selfishness, which served few, instead of all human beings. I am under the great influence of nature, which is serving its creatures selflessly but I am unable to break the strong chains of helplessness of man's progress, which have weakened and captured all natural resources. I don't know whether I can break these chains in the future or not. Like me, man is bound to dance on the tips of man's strong hands of progress.

My soul mate met with the sky, sea, air, earth and sun and told them his grief. All the time nature consoled him. He told nature that time is being executed, and I am also the part of that power and can do nothing. Those who are executing time are able to take power from him, and he is unable to stop them.

The multicolor soul never liked the moderator, who was saying something again, "Time is an illusion. In man's own words, man's history is that of development from insect to reptile to animal to man's shape. During this journey he paved the way to destruction. In a real sense he has executed time, only time was not running. Man was running towards his destruction. No other form of life has a clock, like man, to see the time. Castles of man's imaginations made with the stairs of knowledge are likely to be destroyed soon before the strength of nature. With victory or execution of time, nothing has been changed as man doesn't know what time will write on his grave."

The multicolor soul ignored the moderator and concentrated on his soul mate. His soul mate said, "I am feeling proud, as I took birth in the age when I saw the progress of man at the top. Man had developed from the age when our parents sent us to bring straws of the wild reed grass from the desert land to make a broom to clean the floor and courtyard of the house to save money, because the broom in shop was expensive and our family was unable to buy one. This poverty has persisted in the world, but on the other hand, knowledge has gained power to execute time and everything is ready to go into the grave, including knowledge, as knowledge has dug its own grave."

"The execution of time will give birth to another moon. The moon will be without water. The moon will be without life. The moon will be without air. However, the sun will not prevent its shine from spreading over it and will certainly make room for it among the planets. When life, air and water will be provided the moons, nature will decide, not man. Perhaps when nature decides to provide these facilities to moons, the earth will have become without water and air. Life is deserting his own paradise. Doomsday will definitely give birth to another moon to maintain the balance of the earth and planets and to save the rest of life on the earth. Life that remains after doomsday will enjoy two moons in the sky. Maybe they will again start to gain knowledge," the moderator was saying.

"I am unable to understand what are they saying, my soul mate or moderator. I am unable now to differentiate between them. I am feeling some changes, which I never seen before." The multicolor soul was muttering.

-15-

A very young couple was sitting in a leisurely fashion in the bushes on the sand as in their culture they were not allowed to meet openly but they were in love with each other and they generally meet like this, sometimes in the bushes during day-time and sometimes elsewhere during night-time. They were not angry about their cultural traditions but this was the way they chose to fulfill their wishes.

There was a small hive of yellow wasps on the small jujube tree. Many times the young man killed yellow wasps on the hive with fire or with the poison DDT, out of fear of the yellow wasp's bites. The poison in its bite swallows that part of the body for two to three day. But today he saw a yellow wasp taking birth from the hive and only one other yellow wasp was helping it in this process of coming out from the hive.

"I have heard a soul takes 8,400,000 births. I don't know which birth this wasp is taking." The young man told his lover, pointing out the hive. The young man today had no intention to attack as they were not afraid of the yellow wasp. In fact today, for the first time there was a little bit of sympathy in his mind for the yellow wasp.

"I have heard that after millions of births man's life comes into existence but what does on earth, god knows." His lover said.

They felt somebody passing nearby and they moved apart.

The multicolor soul took birth as a yellow wasp. Now nobody knows each other as soul mates.

End of Novel

2ND Book

2nd

Moon

(Short Stories)

2nd Moon

Bhikhi Raj was very worried about the rumors in heaven and came to Rikhi Raj and asked him, "I have heard that some aliens, beliens, sen men are desperately looking for heaven in order to capture it and they are very near to completing their mission. Any of them may soon capture heaven and take charge of it. They may put us out from heaven. Please tell me the truth about this. I am very worried."

Rikhi Raj smiled at Bhikhi Raj's comments and politely said, "Be calm and sit near me. I will definitely solve the dilemma of your mind."

Bhikhi Raj accordingly sat near Rikhi Raj calmly and without any hesitation.

Rikhi Raj said, "Some part of your talk is very true. There are too many discoveries on some planets and yes, they are trying to discover heaven but they don't know even about the situation of the planet on which they are living."

"What do you want to say?" Bhikhi Raj asked.

"Tell me about whom you want to know first," Rikhi Raj said.

Bikhi Raj, after a little bit of silence, said, "About man who is living on the earth. I have heard they are also very eager to find heaven and to win and control it."

"Alright, I will certainly tell you the reality of the earth and man who lives on it," Rikhi Raj said.

"To know the exact situation, first we have to see their past a little bit. Present progress has not come on the earth for the first time. We should know what happened to their progress in the past. This has not happened for just one time. This has happened so many times on the earth in the past. I will tell you one story out of those many happenings. Once upon a time man in the same way made too much progress like in the present time. Everything from the earth he took out on its surface and the knowledge of man had made the earth too weak, like in the present time. The knowledge of man was prevailing all over and he was ruling on the earth. Men were enjoying the strength of the earth in the same way as they are enjoying it now," Rikhi Raj said.

"What do you mean by the strength of the earth?" Bhikhi Raj asked.

"There are so many sources and things in and outside the earth, which are the strength of the earth. Iron, metal, coal, stones, gases, oil, and so many other things that are helpful and necessary to maintain the balance of the earth, man took out from the earth for his own use recklessly," Rikhi Raj said.

"Man cannot use these things?" Bhikhi Raj asked.

"Man can use them, but man has always used these sources beyond their limits and his needs, uselessly for his ego and greed to fulfill his unnecessary wishes, which have no meaning in the eyes of nature.

"Man doesn't need these things and can enjoy life more comfortably without them. These things are used by man in a reckless way." Rikhi Raj said.

"The same as now, due to a lack of original strength, at that time the earth was unable to maintain its balance. Man at that time also used the sources of earth beyond its limit and raised the danger to the existence of nature and earth," Rikhi Raj said.

"Fear prevailed all over like now with the rumor that the knowledge of man had become so powerful that it would soon take over the possession of heaven as well as other planets in the sky. Everything had become scary in the sky. But it was not true. On the other side, the picture of reality was very different," Rikhi Raj said.

"Actually, this was a tough time for the earth and the circumstances compelled the earth to take action against the knowledge of man to maintain its balance and to regain its strength. Many needful things that give strength and energy were not in the control of the earth and man was using these recklessly," Rikhi Raj continued to explain.

"You know the law of the earth has no effect over the law of nature. Only the law of nature prevails everywhere. The law of earth is made by man to give strength to his ill will and to extend his knowledge. It has nothing to do with the law of nature. Nature has a bigger role to play all the time. The knowledge of man plays a role part of the time only to fulfill his greed. There is no greed or selfishness in the work of nature hence the law of nature prevails everywhere. There is no weight in the eyes of nature of the law of knowledge of man," Rikhi Raj explained

"When the knowledge of man interferes in the bigger role of nature and in its routine work, nature becomes helpless to serve. The knowledge of man left no way to nature except to take action to destroy the knowledge of man. So nature at that time helped the earth to take action to destroy the knowledge of man, which was damaging the processes of earth and nature," Rikhi Raj said.

"The earth is rotating around its axis and also revolves around the sun. The central pole of the earth is the main strength-producing part of the earth. If man will try to control that part of the earth, he will destroy the whole earth. What will happen with the earth at that time, nobody knows. However, that time will never come. The earth will never allow man to do that. The earth will certainly take action before that," Rikhi Raj continued to explain.

"At that time the same thing happened and the earth took action. For the earth to capture all its strength back, it took a hard step. The earth opened its central pole from one side to the other. Everything that was on earth went inside the earth. The inner part of the earth came out. The central pole slowly took its position again in the center of the earth. During this process, the waste part of the earth, which had become the burden on the earth and the earth was unable to bear any more, was separated. Men later on called that part 'moon,' which is still in existence as a stigma on man's behavior and revolves around the earth and with the sunshine looks very beautiful," Rikhi Raj said.

"Man?" Bhikhi Raj asked.

"You are right. How was man saved during this process? Near the equator and on the north and south poles some human beings and other creations were saved. The people were without anything on the earth. There was nothing on the earth. Everything had to start from the first stage. All creation was wandering free without almost any of the sources of the earth, which were hidden under the earth and which the remaining human beings had yet to discover again. Man was also fighting for food. Other creations were also struggling for food. Many dangerous animals took birth on the earth. By eating flesh all the time, they became more dangerous. During this time, man was also struggling. Some of them wrote books about this experience. But those were useless, because nature had totally broken the knowledge of man. There was nothing to reconstruct again on the basis of bookish knowledge. Bookish knowledge became useless for the remaining human beings and with the gap of time they started saying these books were a myth. There was no sign left behind by the doomsday to recognize the bookish knowledge. Man had to reconstruct all again on the fresh knowledge. During the struggle, man had to work for food and shelter," Rikhi Raj said.

"Time is very powerful. During the long history with the struggle, human beings every day gathered new knowledge and again made a big castle of knowledge and, passing through the ages, man again dressed up with the gathered knowledge. However, he has forgotten the consequences of that knowledge, which already brought doomsday in spite of the fact of the written truth. New knowledge and the race of greed again blinded man. He is still not ready to admit the old truth of destruction and ignores it by saying that it is a myth," Rikhi Raj explained further.

"Now again, gathered knowledge is at an extremely powerful stage. The earth is also again ready to take action. This has happened again and again on the earth. Definitely, there will the birth of a second moon soon. Again, the earth will take a deep breath. Again, nature will break the knowledge of man. Again, a part of the earth will be separated. I don't know what

name man will give to that separated piece," Rikhi Raj
finished.

Bhikhi Raj took deep breath and left,
being fully satisfied.

God

Today, Bhikhi Raj was looking very sad. He was silent and not talking as he usually did, when he talked so much.

Rikhi Raj scoffed at him: "What happened, man? Your mouth is shut and your face is dim."

Bhikhi Raj said, "I am very confused today. I don't know any more the solution for my confusion, even."

Rikhi Raj said, "Remove your confusion. But you have to open your mouth with a smile and to explain the reason for your confusion."

Bhikhi Raj said, "Some people have deep faith in god and some are denying the existence of god. People are giving solid reasons and logic to prove their belief. There are wars in the name of god. Some are murdering people in the name of god. In spite of that, they are declaring themselves emissaries of god. Some are saying there is no god. They are giving solid reasons that if there is any god, then why is god ignoring the curse that has fallen on so many people who need his help and have trust in god. Why doesn't god punish the people who are apparently and directly responsible for this curse? They don't think that there is any god. If there is any god, then there must be something to show the justice of god or the existence of god."

"Oh, it means this small cause was the reason of your whole sadness," Rikhi Raj said.

"Do you think it is a small cause? There is mass killing over the earth about these things and you are saying it is a small cause. Man killing man in the name of god or some are killing because they don't believe in god; is it a small cause?" Bhikhi Raj asked.

"It is your weakness that makes you think this, keeping in mind only the needs of man. Only this thing creates confusion in your mind. You always talk about hell in the heaven. You don't think about the whole creation but think about man only. Forget man and then start thinking, and your problem will be solved automatically,' Rikhi Raj said.

"Please tell me something about this clearly. Don't confuse me even more. I am unable to understand your enigma," Bhikhi Raj said. "This is happening nothing like the way in which you are thinking. You are unnecessarily in big confusion," Rikhi Raj said.

"Solve this riddle, please. I am unable to understand your deep philosophy," Rikhi Raj said.

"There is nothing to understand. You are only thinking about man. Think about the whole planet and creation, leaving man behind, and you will be able automatically to understand the whole view of the circle," Rikhi Raj said.

"You are making me more confused. I am totally out of my mind now. What are you saying that I am unable to understand?" Bhikhi Raj said.

"I will tell you in another way to clarify this and to get you away from the influence of man and his belief in god. A beetle eats many ants in the light at the nighttime. Ants also attack the beetle and take a bite of his soft stomach. The beetle writhes and wriggles at the bite and sometimes falls down and is trapped because his wings don't allow him to fly away. The ants again give many bites and ultimately kill the beetle. The beetle's dead or alive body is the favorite food of ants, which they get at the cost of the life of several ants. Now tell me the role of god here. It is there in the routine life and they are fighting in their part of the role and all are busy in their routine and they don't have time to think about god. Otherwise, you tell me what should god do here and in whose favor god should act?" Rikhi Raj said.

Bhikhi Raj could not give any answer and started looking hither and thither.

"I know I have to tell you more to make it clear, then you will be able to understand this mystery," Rikhi Raj said.

"The earth is serving the whole of creation over which it presides. It is not made to serve only man. It is not made to serve only beetles and it is also is not to serve only ants. If any of them uses it for his strength to grab the right of others, then what is the fault of god? The earth is the mother of all and is able to give food to all. If a lion makes a cow his food and kills the cow, then what is the fault of the earth? The earth has given power to each being to defend their rights and they fight for their right too. However, the earth plays no selfish role between the creations over which it has control. If a lion will try to eat or kill a man or a man puts a lion behind bars or kills him, it is between them, the earth has no role to play. We cannot deny the existence of the earth," Rikhi Raj said.

"What you mean? That the earth is god or do you mean that god has made earth?" Bhikhi Raj asked.

"I have not told you anything about god, however, I will tell you about god but don't be eager to know before knowing the reality which I want to tell you first," Rikhi Raj said.

"All right, excuse me, please," Bhikhi Raj said.

"Now you take another example. Air is the breath of every creation on earth. Nobody could be alive without breath or air over the earth. Air provides this breath to its every creation without any selfishness and free of cost and without any discrimination. Air plays its role in nature in many ways. Nobody has control over air. Nobody has power to direct the ways of air. Air is also the main source of nature. Air works tirelessly for nature. The work of nature is not for any specific place or specific species over the earth such as man. Man's work is not like air. Man is also one of the species. About what he does, air has no concern. Air does not work with the cooperation of any species. All depends on air but air does not depend on anybody," Rikhi Raj said.

Bhikhi Raj was listening seriously. Now he was feeling that he was gaining vast knowledge from Rikhi Raj.

"You can take another example. The sun shines for all. The work of the sun is not for any specific part of the earth or any species on the earth. The work of the sun is not even limited to the earth. His work and role is too vast and cannot be counted. The great work of the sun is beyond the reach of man. Man and his work are too small before the work of nature and its main sources like the sun. The shining of the sun gives life to the whole of creation upon the earth jointly and together. Life has no meaning without the sunshine. The sun is the major part of life but life gives nothing to the sun or its shining and has no role in the sun's work," Rikhi Raj said.

"Now I am beginning to understand a little bit," Bhikhi Raj said.

"For more clarification I will tell you the whole process. The sky is unlimited and cannot be measured. You cannot estimate the length and width of the sky. You cannot estimate how many earths and suns are in the sky. The sky is not in control of anybody. Man comes and goes over the earth and other species come and go over the earth. From that history, nobody knows how much time this has been going on but no one on the earth could come to know about the greatness of the sky. Nothing is in the hands of man or anybody else to know or estimate the greatness of the sky. It is beyond the knowledge of anyone. The sky is great and its power cannot be estimated," Rikhi Raj said.

"Now I am going to be very clear," Bhikhi Raj said.

"One thing more I want to tell you. Upon the earth, water also has the major role to nourish life. Without water, there is no meaning of life. The body cannot exist without water. Water makes life possible for each body, even of man's body, but a body cannot make water. Some people upon earth believe the earth is a mother and water is a father of the body or life. There is nothing wrong in their belief. It is one of the truths and truth cannot be denied. How water plays a role upon the earth and how it nourishes the life without selfness, nobody knows and nobody can play such a huge role upon the earth. Man is too small before the working of nature. Man or his beliefs or his working has nothing to do with the work of nature. What man does or what he does with other men or other species or other species with man—nature has nothing to do with that. Nature is busy in its routine work and its work is selfless."

"I fully understand the reality now and my confusion has almost been removed," Bhikhi Raj said.

"I also want to conclude now. Now if you, man, or any species addresses this working of nature and working of its sources jointly as the work of god or gives the name of this joint work the definition of being god's work, then what is wrong in it? Man has no power to play such a big role. Otherwise, what is the age of man? He will die in one hundred years. His knowledge on earth will also not stay forever. What man says or what man is doing is man's course and not nature or god's course. Man having an atheistic or theistic nature has nothing to do with it. Who kills whom nature has nothing to do with it. It is not the course of nature," Rikhi Raj said.

"Now the answer to your question is very clear. Some persons on earth have worked very neatly and cleanly for the other human beings like the work of nature. Their selfless work has given them the respect of others and human beings have regarded them like the work of nature. Their sacrifice made them stronger. People regarded them like gods. Man is the victim of man. If they are confusing each other for their selfishness, then what do god or nature have to do? They have to clarify themselves. If they want to survive, they have to struggle against any cruelty. Nature or God have nothing to do with that. Nature or god will never stop anybody from doing their job, whether good or bad," Rikhi Raj said.

Bhikhi Raj was looking very pleased now.

War

His upside was down. He was shaking his legs continuously, which showed that he was still alive. Actually, he was squirming. The upper side, which was down now, had become white with dust on the earth. It looked like before accepting defeat he struggled too much to maintain himself in the form. He must have struggled to remain upright but must have failed and become fatigued. He was shaking his legs in pain. Actually, it looked like he had lost the war and was taking his last breath.

With the view to save him, I put him upright with the toe of a shoe. I don't know whether he got a new beam of light for life from this or he was making his last effort to save himself from the pain, but he was running very fast. Sometimes with his left or sometimes the right, with his wings he was thrashing the earth. So many ants fell away when I put him upright with a toe. With his thrashing wings and his fast running, some of the ants were going away from his body. Some of them he was eating in haste. My sympathy was shifting from his safety towards the deaths of so many ants. I was feeling that I have taken the life of so many ants by saving him. But I did not dare to turn him over again and to hand him over to the ants. I don't know the reason; I just decided to see this war as it is.

It looked like the beetle's hand was up now in this war. He ate so many ants, which he was able to entangle. This was in spite of the fact that he was very much injured and a big part of the back side of his body had been eaten by the ants and damaged completely. In other words, he was on the death bed at this time but at the same time, it was his last chance to save himself. I was not clear that he was saving himself from pain or saving his life. He did separate successfully himself from almost all of the ants. Not all left him free but most of the ants ran away from him. I think the ants were of the view to lessen the damage to their lives.

Now, his major weakness was that he was not able to fly. The back side of his body was injured completely and his wings were also damaged and were not in balance. As a consequence, he was not able to walk freely like other beetles. He was not able to close his thick two wings like other beetles. He was in a very bad condition. He was not able to return to his original form. In reality, even if he might have been able to save himself from the attack of the ants, it was certain that he could not remain alive because he was injured very badly. It looked like his death was ultimately certain.

However, he went on. The ants could not run as fast as he could. I think the ants knew his fate. They were aware about his condition. Where could he go? When he was falling on his back and many ants were on his body and were eating the injured part of his body freely, then his condition was become very pitiable. Now it was very clear that he couldn't go away from their range.

A few ants were still chasing him. When any ant bit at the injured part of his back, he fluttered noticeably. But if any ant came before him, he ate it in one swoop. The war was going on. I had to go check another side but at the same place very near to it.

I came back after a few minutes. I looked for him to see his condition. Away from that site, he had again fallen upside down and the ants were dominating him and freely attacking and eating his body. He was shaking his legs but this time it look like he was in even worse condition. But it looked like he was still alive and needed help to save his life.

It was not totally fair for me to be one sided. If I put him upright, he would have to eat more ants and my decision to help him would adversely affect the right of ants because so many ants had already sacrificed their life. In spite of this double-mindedness, I put him upright again with the toe of my shoe. This time he was in a very bad plight. It looked like the ants were aware of his condition. He was not very able to save himself from their attacks.

I had to go. I was on duty. I could not watch this war any more. It came to my mind that I could save the beetle from a painful death by being eaten alive by the ants; I could put him on some safe place but I could not find that place. The ants were able to reach anywhere. I was in a hurry, too.

Today was the second day. The valiant soldier of the war has become a martyr. Ants were ruling over his dead body. It looked like an ant hill. I don't know how tasty that food was for them, because they got it after sacrificing the life of so many ants. For the death of some ants, I was entirely responsible. The martyr had fallen dead in the same upside down position but there was no movement in his body. Now, there was no embargo on ants to go on his body and to eat.

I become a little bit ambitious and interested to know more about this phenomenon and about the war between beetles and ants and a little bit about their life. I saw one more beetle on the ground, which appeared to be dead. However, when I touched this one a little bit with the toe of my shoe, he ran away fast and went under a pipe lying there. Some distance way was another beetle. I watched his activities for a while because now I was aware that the beetle was lying there intentionally without any movements to confuse the ants, which were under his attack as food.

Two or three ants were wandering around him. It was difficult to identify his back or front side as he had closed his wings. This was the cleverness of beetles as part of their strategy in their war with the ants, which I noted today. Whenever an ant comes into the beetle's range to his front side, he brings his head ahead and eats the ant and takes the same position as if nothing had happened. Even a nearby ant knows nothing about what just happened there.

Actually, this was war. Those were spy ants. They were unable to understand the beetle's front or back side. They sacrificed the life of one or two ants to identify his back or front side. Then they attacked from the back side. At the back side, under the thick and hard wings, was the very soft stomach of the beetle. The ants took a bite at the stomach very cleverly and quickly under the attacking strategy. At once, the beetle squirmed with pain and flew into the air to escape the ants but struck an electric pole and fell upside down on the ground. Many of the ants were apart from his body but he was trying to put himself upright very fast to save himself from any further attack of the ants but was unable to set himself upright. Many beetles were there as well as lots of ants and their war to kill each other for food was continuously going on.

I discarded the idea to understand the war any more. Before I could come to know the reasons of their victory or defeat or know more about their routine life and before this, the wars of man came to caught me, I left the place.

High Spirit

Resting his head on both front legs, looking relaxed, he was in deep thought. It looked as if he had accepted complete defeat. He remembered those days when all called him a king. He never ate contaminated food when he was free, never ate the flesh of the dead. He always hunted and ate fresh meat. Even other animals took hope from scavenging the flesh that he left and that they ate.

He was lying helpless in the deep ditch, which had been made to hold him with awareness of his power. These days, however, his power could not take him out of this ditch. Onlookers wandering up were laughing at his helplessness, although they sometimes became frightened by taking just one glimpse at him.

When they put him down in the ditch for the first time, then they first thought about his food. They were very well able to kill him but he should be thankful forever for their kindness. Now there was no need for him to go to hunt anymore. Now he will be served with fresh meat. For many days, he didn't accept this meat. He was a lion, a king of the jungle. Whenever man wants to show himself as powerful, he compares himself with the lion. But now his helplessness does not allow him to remain depends upon his bravery. To kill self by hunger is not a wise decision. He has to give up his persistence. He has to accept contaminated meat. He has to change his habits according to time. Although his sense of honor has not died yet, he hasn't given up roaring. Man still becomes overcome with fear when he roars and the lion started living only on this contentment. This was in spite of the fact that his doors were closed with thick rods due to man's own fear, which still reminds everyone of the power and existence of a lion.

With the passing of time, when he heard the cry of man in the same way as he cried in his own helplessness, he felt a strange type of satisfaction. Slowly, slowly he felt that the people watching him, who are wandering freely, actually are tied in a strong system and are more helpless than him and suffer under a greater psychological atrocity than he. Animals, which don't make strict laws and rules and depend upon the natural law, enjoy an easier life. Man is confined in the chains of his own laws, just as the system took the lion and shut him in the box and he is helpless. Beyond doubt, death is not away from him who dared to fight against this system, but for the one who tries to look out of this system, the consequences are more severe.

He came to know that god took birth on earth several times to correct all its shortcomings but every time, god was killed miserably. Sometimes he was hanged from a rope, sometimes hanged on a cross, sometimes forced to sit on hot iron plate, sometimes put in hot water in a large kettle to die and so many other ways by which people can be killed. Man became atheistic too but all the time he was unsuccessful to achieve anything.

Man was not the only victim of atrocities but all of creation and nature became the victim. There is no way of going behind to fight against these people. All the sources of nature are in the possession of a few hands and rest of creation must live in fear of these hands. There no inch of land on which there is no stamp of ownership. Everybody is bound to accept their right of ownership. No one can live independent without accepting the effect of their possession. The lion heard from man that it is his sense of honor that put him behind bars otherwise, if he would have given the proof of loyalty like us and dogs, he would also be able to wander free like us. He heard from man that he should not worry about it because those people who don't obey their order, they were put behind bars like him also and they cannot come out until they become gentle and were forced by fear to obey any orders they were given.

Man is bound to live in the environment as though he were contained in packed boxes and all natural resources are also packed in the boxes and are reshaped under the stamp of the manipulator and man is bound to eat contaminated food. Sickness becomes common so medicines are necessary, and the average person must take these every day with food. If you are unable to sleep, don't worry about that, they will give you the means to fall asleep. You can die without any pain; however, only if you have expensive insurance or a lot of money in your pocket. To reduce the effect of poison, other poisons are available. Now harmful sugars will be produced by your body.

There is no way that man could escape from this environment. Everywhere there is the stamp of the trademark of those who benefit from this unnatural situation. They have cut off the relation of common man from nature as they are ruling over nature and the rest of creation is under their control. Man can purchase things made from natural sources with currency made by them of paper, which he can earn at the cost of his blood and bones. A loan is also available everywhere, which you can get by mortgaging the rest of your life. You can get even more money by mortgaging the life of your unborn child. All other ways are closed and you are bound to die in this environment.

Natural sources of fruits and vegetables ripen with the aid of poisoned chemicals. Otherwise, nobody allows those to ripen on the plant. This is done in order to make sure these fruits and vegetables ripen in the hands of those who farm only for profit. In order to get all possible profit from people's health, they have made strict law and rules.

He heard from man that many men are born like him, roared like him and sometimes they also refuse to accept defeat but the consequences of those were even more terrible than what he suffered. He has not that kind of wisdom that he can convince them with arguments. They become the victim of atrocity because atrocities make others passive with the use of guns and bayonets. Those who don't become silent by the threats of guns and bayonet, they were forced to decay in the hell. Man is not only packed in boxes, he is being guarded shadow of guns and bayonets. It is not possible to raise one's head, nor is it acceptable, either.

By hearing these things about people, the lion was very satisfied. By sharing his grief, he reduces it. He was very peaceful now. Now he was not irritated to see people around him. He considered them as existing under coerced circumstances like him; they looked free but there was slavery in their minds in several ways and at every step. The lion was unable to understand the economic, social, or slavery aspects of human systems. His slavery was slavery and freedom should be freedom.

At first, the lion was irritated by the sparrows, which seemed too much like man. However, when he heard about the grief of man all his grudge disappeared. He was very surprised by hearing the song of the sparrow full of courage in the cage. Till today, he believed himself a king but today by taking a new type of inspiration, he was feeling very happy.

He came to know from the sparrow, what is the real sense of freedom? The sparrow challenged the man, that he has provided all types food which was not even available when the sparrow was free, but if the man has confidence in his own services, then once, let him open the window of my cage and see how much I trust him and see how much I have become the slave of his obligations. I will never turn back to see his face again. The sparrow challenged that if it will have to die in this cage, it will die with the hope of freedom. If man really had any pride of his delicious meals and other produces, he should have not put it in this cage. He tries to make me a slave of his production but I am confident that he will never succeed.

When the lion heard the stories about man, the sparrow challenged the lion at that time too. The sparrow said, "Man is the slave of his needs and does not nourish the desire of freedom in his heart and so men die in envy of each other. They have polluted the air and water and that day is not away when man will go out with a cylinder of oxygen on his stomach."

Maybe the lion did not care about the sparrow's talk when she was singing the songs but now with the mixed influence of the sparrow's high spirits and on the poorness of man and his self helplessness, he was deep in thought, keeping his head on both front legs, looking relaxed, and keeping helplessness in mind.

Broken Star of Heaven

By leaving your soma, you cannot get relief. The malevolence of the unfulfilled desires of your body that have been left behind in your soul, which has not been separated totally, does not allow you to get relief and your soul has strayed fruitlessly. I also am a strayed soul whose body has been lost and in the unfulfilled situation am feeling the presence of earth where my body was lastly separated from my soul. On my straying situation the sky is laughing, and the elements of air, water, fire, and earth, which were part of my body and now are separated from me, are smiling.

Otherwise, there is no deficiency of earths in the sky. Any sights you will see in the sky are earths and earths are shining like stars. I never saw the importance of the sky in the soma like this. I never felt the cosmos like this. That which I am feeling right now has decreased my overly great attachment with the earth. This will clear the way towards salvation because this attachment is the main element that has not separated from me yet.

There may be another reason that my soul has not been freed yet. I don't know what will happen after salvation. After salvation, I will get another body or not. If I will get a body, then I will learn on which earth, when and which type of body my soul will get. This straying situation is such a hard truth, which I never thought about or realized when I was living in soma.

In the soma, I always thought that the earth is my mother and the water is my father and I worshiped them. My belief was that life on the planet, of living in the days, nights, and rotating in the seasons, is the truth. At the time of leaving soma, my soul felt no grief but now at the time of embracing truth, prior attachments are becoming heavy. The door of salvation was not looking more beautiful than the influence of prior attachments.

You cannot explain the beauty of mother. You cannot praise father. You cannot expect a more beautiful heaven than the earth. The heaven that you are enjoying is in the lap of nature when you are in the soma.

Soma has only one heaven but a soul has many things above that. This satisfies me. Attachments also satisfy me. Otherwise, the soul has no satisfaction. A soul has no thirst. However, when it enters into the soma, thirst also takes birth, which, after leaving soma, becomes a fear of wandering.

The blessings enjoyed on earth have become the reason to wander after leaving soma. Attachments are realizing the existence of the sky.

The sky is full of suns and earths but attachments are forcing me to have realizations about the life spent on one earth.

The sky has a reason for laughing at me. The apprehension I am feeling is the routine work of the sky. The stars are broken in the sky all the time and there is no effect on the sky of these incidents.

My affection for the earth is due to the recently enjoyed life I experienced on earth. I have seen that man under the flag of progress based on his knowledge is continuously trying to ruin the earth. In his pride, he is declaring the victory over nature. On the other hand, he is shedding crocodile tears over the curse of the earth. He is trying to become the father of Mother Nature. In his ignorance, he still does not understand that he is not the driver of Mother Nature. He is unable to understand the power of his father, which is water. He does not understand that he does not make day and night. He does not understand the power of the sky. He does not understand that he cannot run away from Mother Nature. He is unable to understand his own ignorance.

It is believed that when a pewit keeps at night, it keeps its legs towards the sky due to its fears that the sky will fall on it while it sleeps. It is a very good example to understand the reality but the confusion of man is bigger than that of the pewit. How ridiculous it is that after putting the earth's existence in danger, he shows himself to be awakened like the pewit. He does this by talking about global warming and talking idiotically about how to cure and control it instead of feeling ashamed for what he has done.

It is not global warming only, but the inner side of the earth has become cold and weak and its strength is being lost day by day. Man has used the natural sources blindly for his absurd progress based on his knowledge. Man has left no way to save the earth except to ruin it. Earth had fathomless, immeasurable and countless power to maintain its balance but man, in the name of progress, has taken out all natural sources from the earth in the form of oil, iron, metal, mineral, gases, and everything that was his need without understanding the consequences . By trading the natural sources at large for his selfish purposes, he has interfered in the routine and damaged the balance of the earth. Man has polluted the earth, air, water, and sky with dangerous and poisonous gases, viruses, and bombs. In this way, he has spread the fire everywhere to capture the natural sources. In the fire of progress, the earth has been scorched.

Traders of progress have not stopped on only this. The clever thought of such a man has taken another step ahead. Now he is searching for more earths in the sky to capture. Frightened by looking at the broken star of the earth, he is in search of other planets as he wants to make sure to be able to stay there in the case of an emergency. He wants salvation from the fear of the earth's end and is eager to tell about his new progress.

The Sky is laughing at me because he knows that it is the routine work in it and my fear is useless.

It looks like the end of my wandering has come. I am getting another soma. I am taking birth on the same broken star.

Affection has also taken birth in me. The same broken star looks to me like heaven. Love for heaven is again becoming my weakness.

Hounds behind the Hare

In a happy mood, Bhikhi Raj tried to tell Rikhi Raj, "When I was in the body of a man on the earth, I saw a very interesting race between the hounds and a hare. The hare would not have any other way or trick to save his life except to win the race from the hounds behind him. I always beg from god the victory of the hare in that race. But god has not always accepted my prayer. The hare becomes the victim and his waist comes in the mouth of one of the hounds and the hound threw back up from his neck with full strength so that it tore the stomach of the hare badly and it became unable to run forever. Hounds actually don't hunt for themselves like lions. A lion doesn't hunt for others. He hunts for himself. That is another reason there are no leftovers when the lion eats other animals. But a lion doesn't hunt like hound, which does so for his owner to whom he is faithful. Lion hunts for his hunger and a hound does so for his owner's hunger. That is another reason why the owner takes care of his faithful dog and gives him some share of the prey."

"I am unable to understand what you want to tell me by telling this story," Rikhi Raj said.

Bhikhi Raj smiled and said, "I can tell you nothing. If I say something to you, it means I want to know something from you and you know my weakness."

"Don't try to flatter me. Come to the real point. What do you want to know?" Rikhi Raj said, becoming serious. "I have heard from you that man has made the earth a cage for man. Now the earth is a cage for all men. Laws of countries don't allow man to move freely. Man cannot ignore the boundaries of countries like birds can. Man cannot go anywhere like animals without caring about the boundaries. If a dog crosses the border of a country, it is not illegal but if a man crosses the border, it is illegal. Everywhere there is the stamp of ownership of someone and every man is bound to admit that by law. Man before man to save his life is running fast like the hare to win the race of life," Bhikhi Raj said and both smiled by looking at each other.

"I think you want to say something else but are saying it this way or that way but not directly. Say clearly what you want to say," Rikhi Raj said.

"When I was on earth, my father told me many things about the ages of man. He told me that there were four ages: Satyug (golden age), treta (silver), duaper (bronze), and kali yuga (Iron). He told me that we are now living in the kali yuga (iron) age. He told me many stories of all the ages. He sometimes said there will be an end of this age too and a satyug (golden) age will come again. Is this world going to end under that system? I want to know the reality," Bhikhi Raj said, directly touching the point in his mind.

"You are very smart. You brought the matter from where to where. I do not agree with your father fully. But the stories that are made by the thinkers and philosophers cannot be thrown away without any reason. Sometimes thinkers and philosophers don't say everything directly. Sometime they speak according to the time that is prevailing and mention many things that have yet to happen in the future. So sometimes, the meaning changes with the time. Man sometimes takes the meaning of everything according to the circumstances of that time or in that situation. Sometimes they take the meaning of the philosophy according to their own selfish interest. Sometimes some people don't want knowingly to come to the truth due to greed and interests in trade. In the same way, man sometimes explains his life by dividing it into four parts. He considers his age to be one hundred years. The first twenty-five years he considers as childhood and for education. The next twenty-five years are for young age to enjoy the life and to give birth to children and to raise them. The next twenty-five years are to sum up the responsibilities. The next twenty-five years are those of old age. But this is not a strict and fast formula. This does not happen in life all the time. But in normal life, generally this happens in that way. In the same way, the ages are not the strict and fast formula," Rikhi Raj said.

"Concerning how the moon came into existence. It is science not myth or automatically adjudged by anybody and there is no record of these things but it has been proven that the moon was part of the earth. Nothing about this is easy to tell because time played this game, which is not understandable in the era of time. Time consists of present, past, and future. The past is your history to research but the present you can define in any way before you have seen what happens at that time, and how the future will give relief or grief, nobody knows," Rikhi Raj continued to tell.

"Maybe man thinks that whenever doomsday comes these four stages take the path again on the earth after that. The meaning sometimes automatically gives shape to the time. After the doomsday, only a few creations are left behind and naturally there are enough resources on earth and there is nothing to divide. Life must have to depend upon principles and you can name it satyug or golden age. Creation's expansion or population's increase must have converted the color from golden to silver or treta. According to Hindu mythology or philosophy, the third stage came first instead of the second. It shows that lawlessness must have taken the bad shape, which is why the rule of law, which is called duaper or bronze, must have come later. The bad shape of law must have frustrated man so he must have given the name to his bad laws, which are away from nature and for his deeds, which must have given birth to vanity, that is, the iron age or kali yuga. This is all fixed by man according to his own situation and thoughts. There is nothing fixed by nature. Nature has no role in this. Nature is playing its role on its own. Nature controls everything, not man. Nature has a bigger role to play. The names that man gives to ages or time, has nothing to do with nature. Your father must have shared with you the knowledge that was prevailing at that time. However, it is very clear that the last decision of nature is doomsday. Satyug or the golden stage never comes automatically. There must be the end of all yugas or ages," Rikhi Raj explained.

"And the second moon will take birth!" Bhikhi Raj repeated the old words of Rikhi Raj and both laughed boisterously.

"If some people are left behind on this second moon, then what will happen?" Bhikhi Raj gave a taunting remark again.

"Man has prepared himself to stay on the moon. He can stay there during doomsday," Rikhi Raj also said in the same way in reply.

"On coming back to the earth, he will have left his coat and pants there. On the earth, he will never find anything like that again," Rikhi Raj made another joke and both laughed again.

Waves Blue Water Green

My right and left sides are seen from my front side. Maybe the meaning will be different if seen from the back side. I never saw this carefully. Otherwise, there is no difference. I will be the same with different meanings. Both parts are mine. To find out the real meanings of right and left, they searched my head many times. This was their confusion and they did this to satisfy themselves. I never found anything in myself anything like left or right.

They don't allow me to speak. They say that they know everything. I don't know what they will steal from my mind. When they get an award for their thinking, then I come to know that this thought was mine. I think they have machinery and equipment to search my head. They have developed such systems to search heads like this. It is not me alone who is being searched. There are so many other heads of different qualities and thinking that are being searched like mine. There are scholars who used our materials taken from these searches to which they call their specialists. They believe them, not us. They give the name right or left.

For millions, billions year I have had the same face, I have said the same thing and all the time was put into the same grave. I have never found anything left or right in myself.

There is no color of water. But it looks like the color of water was blue. Waves that were coming towards the shore appear to be green. I don't know whether this was due to the air, sunshine, or earth below the water or it was the illusion of my eyes but what I was seeing in front of me was real.

People generally say that nobody can count the hairs of the head. I also cannot count the hairs on my chest. I even was unable to count the people sitting on the shore of the sea. I was unable to count the birds flying in the sky. But how many people are living on earth has been counted for many years. Birds were not accounted for. Maybe birds could not be counted or there was no need. Those animals that needed to be counted were counted and the rest were unaccounted but estimates of their numbers were made.

Calculations of how many people live on the earth always change. My right and left sides always remain busy to do something. They remain busy through bombs being used to capture my chest. They declare that they are sowing the seeds of freedom and this is the wish of our people. So they changed the meanings of so many lives into death. They declare that death is the truth of life. They compelled people to admit this truth.

When they searched my head for the first time, I saw that they were looking for something left. Actually, there was no difference of right or left for them but to erode my thinking as a whole, it was necessary for them to put me in some sheaf of thinking. Maybe they want to put my thinking in the bag of left. Maybe they were willing to make my thinking the feed of some animal. With this, they can give the power to the body of animals and can use the waste of animals to grow their crops. However, all the time some part of my thinking always drops from the sheaf and the remainder of my thinking lies there neglected like stalks left over in the feeding trough by the animals. On the failure of efforts by the rights, the lefts raised their flag. They were in possession of a big part of my body. With their distinguished thinking, they were also controlling a big part of people's power. This distinguished thinking then divided into so many colors, which became self-rivals. They were shaking their long hand with the long hand of rights under water. They were in a hurry to stamp me in some colorful thinking. It was not easy to stamp me on the back side, so they were willing to stamp me on the front side.

The effect of bombs of the left will not be different from those of the right as with the same slogan of freedom, the same drenching of my chest with blood, the same heap of dead bodies, and the same persons counting the dead bodies. Freedom spread by the left and right has the same stink.

The left will also try to search my head to bury my thinking under their ideology. Because the sound of the bombs will be the same, so the right and left could take action jointly. I am only I am. I am not right or left. Mass killings over my chest are also mine. On which part this mass killing happens is under the control of the power of the right and left. The broken part of my chest is that ox of the oil press whose limits have been fixed but his journey will end only with his death. So my body cannot be divided into left or right. Left thinking will also do the same job to crush my thinking as to prevail their way of thinking? My ears cannot tell the difference of the sound of bombs of the right or left, my eyes cannot tell the difference of the blood flowing on my chest. The voice of humanity arrested by attractive thinking reaches to my ears and eyes.

Whatever the thinking is, the bombs of wisdom will fall on my chest. It has become the destiny of my chest to bear all this. It is difficult to raise the voice of humanity now because the rivers of blood on my chest are enough to flood that voice. Now my chest only knows the language of bombs. The meaning of this is that life has become to live under their terror.

In spite of all this, I don't know why my body is looking so vast like that mother whose son has just been finished before her eyes or like the earth on which this universal play is going on. I don't know what will be the result of more searching from my head? My chest is already bearing calmly the stink of bombs, rivers of blood, beastliness of thoughts, and silence of innocents.

Conversation

Bhikhi Raj again today felt exhausted, so he went to Rikhi Raj in heaven and said, "Man thinks himself so learned, knowledgeable, and enlightened but why he doesn't he change his behavior towards nature before the doomsday? He can surrender himself before nature and can help nature to cure itself as it is the only way and without this, there is no other way. If the whole creation is at stake, then why doesn't he think about this to surrender and help nature before a dreadful day happens?"

Rikhi Raj consoled Bhikhi Raj and said, "You are being unnecessarily worried about man and his doings on the earth and his unfair behavior towards nature. We are not very much concerned about his doings."

Bhikhi Raj said, "I am not worried about man but I am worried about the earth. You have said that the earth has to take a decision under compulsion to maintain its balance to break the knowledge of man, which has become dangerous to the existence of the earth. But if he corrects himself and

213

tries to use his knowledge properly then learned man can stop this unexpected occurrence. What harm is in that?"

Rikhi Raj calmed Bhikhi Raj again and said, "It is not possible now. Man has established such a culture that he cannot go backwards from that at this stage. He can't go to an early age. It is not possible for him to go behind. Now everything is over. Man has knowledge, power, and resources to take out the sources of the earth but he has no power or knowledge to return the same to the earth. Otherwise, he will never agree to give up the use and comfort he has gotten from his discovered miracles, the use of which has become habitual to him, making him its slave, which he thinks is the gift of his science. He will accept to die but will not be ready to give up his progress. It is totally not possible for man now at this stage."

Rikhi Raj continued to say, "You can take the example of anything. Man has developed cars, airplanes, computers, phones, and many other things and uses these things for his routine work. Can he get rid of these comfortable things for the sake of the earth and nature, can he give up those luxurious and excellent facilities that have become his life style? In many countries, these things have become the symbol of show, while in some countries, they are needed for his daily routine and in some countries, they are just a craze. Land and the houses on it have been fully captured by the capitalists and common man has been compelled to live on rent or pay installments. The car is not only the symbol but has become the legs of man in the attractive world in some developed countries. Without legs, a man has no value, which means that without a car, you have no value. A car needs not only too much iron but to run it needs too much gasoline. A car will be the most enticing thing of this age of the earth. There are airplanes but man can't show his pride running with him in one like a car. In the air, nobody can see his pride. But airplanes will also be the miracles of the advanced age, which means thousands of years after doomsday and after the destruction of the knowledge of man when the generations of man will read the stories of their forefather. Some will say that it is a myth and some will believe it to be a truth. After a gap of time, man will again gather knowledge and create a practical life but they will ask for the proof, which will not be available until the next progress like this one, after which destruction will again due."

"Earth is the main source of food, which living beings need to live. It was good if man would have used it for good and healthy food but man could not keep himself limited to such simple food. He discovered many such things for his luxurious lifestyle and his knowledge put him on the mountain of greed. He forgets his limits. Now nobody knows when the mountain of greed will slide towards the sea of destruction. Man's greed is ruling over man and the earth. Even if man decides to return to a simpler state, his greed will never allow him to do that. There is a limit of everything, even of the earth but there is no limit of the greed of man. Every man sleeps after eating his food and wakes up to eat more food and needs nothing more than this. However, you can see his greed everywhere on earth, which has no meaning for nature except the danger it creates for the earth," Rikhi Raj further said.

"Man's pride will never be able to win the earth. Nobody knows how many generations have come on earth and tried to win the earth but at the end, they were not more than a handful of ash or earth and their pride is buried in the earth forever. There is nothing alive on earth except the pride in some books of history. Man's pride is his history. In the past, his pride gave the coming generations a path towards destructions. This is his only achievement that his pride has earned so far," Rikhi Raj continued to explain.

"The earth is only a friend of life till its end. It gives a guarantee of food to all of life. Beyond that, all is greed. There can be no guarantee of greed. The possessions of the greedy are their graves. They are the owners of their graves. They are ruling over their graves. They are ruling over their greed by eating food, which is the gift of nature. They are not thankful to nature. They think that they are ruling over nature with their science. This false notion is the product of their greed." Rikhi Raj said and continued to explain.

"It is man who is claiming his victory with the progress of science, but this has brought him to that turning point from which he cannot go back even if he wishes to. He is accused in the court of nature. He has been declared guilty and only the sentence is waiting. Nature will decide his case soon." Rikhi Raj concluded.

Bhikhi Raj was feeling half sleepy. He was not even interested to listen any more. Rikhi Raj knows his habit and left the site to do some other work.

Doleful

The farmers have taken away the peanut crop from their sand hill farms. These farms were capable of giving only one crop. The rest of the time, the fields remain lying vacant. Hence, groups of cranes have come here to eat the peanuts left behind in the fields. When man fixes the birds in an era of friendship and enemy birds, it may be that he considers these merits only. It was their urbaneness to come after the crops had been taken away from the fields. They must have waited till the crops were ripe and must have given the time to the farmers to take the crops away to their homes. That's why the farmers put them into the friendship era. They have migrated here after flying a long distance. But they were not even feeling tired. Rather, it was the inner happiness of having reached the destination. At the time of digging from the sand, some ground nuts still stayed in the land. It is very easy for cranes to find ground nuts in the sand farms. It is one of the favorite feed of cranes, to which cranes are compelled to come after having flown such a long distance. It also creates a friendly relation with the farmers, as the cranes don't harm the crops. The farmers also accept them, as they are not harmful in any way. Cranes are obliged to the farmers as they become able to get this feed only due to the farmers. It is a very good example of dependency on each other without harming anybody to fulfill their selfness.

Cranes are a very well-disciplined type of bird that lives in big families, which gives a glimpse of a social society. Spy cranes, without caring for the boundaries of countries, are able to find the best feed and its availability by quantity and quality anywhere in the world, moreover according to the season and weather. Cranes are habituated to cold weather, having the quality to advance and control their actions jointly or individually, are very alert and vigilant, very dedicated to each other by heart and the owners of many other characteristics by virtue of which they have the knowledge to live in the social society. To find the feed from any corner of the world in big quantity for a long period and to take decision to go to that place is indicative of their wisdom, capability, competency, and intelligence. Also, the fact that cranes are not harmful to any other species and live in a friendly atmosphere gives another impression of their good nature.

Their distinguished recognition is also due to their way of living as a couple. You can see them at odds only at times of big trouble or due to their offspring. Each spouse cannot live beyond the time his or her companion dies. It is the habit of cranes to live in the social society and to do everything together to find their livelihood, but their society is not more burdened than other species.

There is no special friendship with the farmers but they find this feed due to the farmers so they were being especially thankful of farmers again and again in their mind. After crossing many countries and making a long journey, they reached here. They were well aware of the poverty of the local people here. They are able to judge the condition of the people from observing the hustle and bustle shown on their faces but without knowing the reasons. Otherwise, they have no concern about this. One difference between how they think about their own kind and that of humans is that the spies of man tend not to be selfless like those of the cranes, who work for the entire community; otherwise, there is no reason for the difference of living standard between man to man.

None of these considerations occurred to the cranes as they saw the disappointed and hopeless faces of those whose left-behind grain they have come to eat. They were unable to do anything for them but before coming to the fields and after taking their feed, they sang songs in happiness for them. They sang songs for their children and when their children also tried to repeat their voice, the cranes understood that they want to listen to more songs and they gave a reply in the same way and sang songs. They give, get, and share all the little moments of happiness.

Today was their first day. They were looking very happy for having reached the destination. They were being thankful of those men along with nature, through whose actions again and again they got this feed.

They were not aware about the ups and downs of man's life but today, due to their ability to find the feed anywhere in the world and having reached the destination freely by flying, they felt pride. Man, however, was unable to fly like them and they were sympathetic with man. They were also aware about airplanes, people riding and coming down. They were very frightened by airplanes and the people sitting in them, but could see that the farmers were not acting in scary ways. They realized not all people are scary.

The people working in the fields were not too scary but nevertheless, the cranes were very vigilant about their safety. When they were busy searching for feed in the fields, some of the cranes remained on watch all around. They were very thankful to man as they got feed due to them but they didn't want to pay for the obligation by putting their safety in danger.

They have no concern about how men deal with other men, what men do with men. They also have no concern about how man goes to sleep hungry and how this feed is produced. But they were wondering why man, who was able to produce such a big quantity of feed, is so worried about his future. They were satisfied by eating these ground nuts left behind by the farmers and they do not worry about the future. They depend upon the treasures of nature.

They were surprised that man has big houses to store the feed, and that the farmers have land and other equipment to grow crops. They could see fields of crops too all around but wondered why man is still dissatisfied and disappointed?

In spite of fact that they were thankful to the farmers for having left behind this feed, which was their favorite, they were very vigilant. To be vigilant is part of their nature.

Suddenly, the cranes were surprised by the sound of firing, which came from hidden places that they could not guess in advance. They only could hear the sound of fire. They flew all together. One wounded crane fell to the ground, which they could see after flying over. She was squirming on the ground drenched with blood. They were unable to do anything for their wounded companion. They considered they had better leave the spot. One crane became separated from the other cranes again and again and then he joined them again. It had been looking for the wounded crane, which was its spouse. Although all the cranes were bewailing their fate, the grief of this crane seemed to be even greater than all the others.

At night it came back to see that spot. It saw the blood and wings lying hither and thither, which told the whole story. Outside the village in a pit, it saw more wings, which clarified the whole truth. No one can imagine its grief at this time. It came back to be with the other cranes.

It was the time to decide. All the cranes were together in grief and they decided not to stay here anymore. They decided to ignore the tasty feed and to fly away. They prayed for the departed souls by saying that the good of the men had been soured by the sacrifice of their companion.

When they were going back, some more flew off, and the cranes met them on the way without asking any further questions. By the time they reached the destination, the cranes had been flying too long. The spouse of the dead crane could not fly. No other crane heard about it further.

Peace

Today all was looking peaceful. It was drizzling outside and the atmosphere was very pleasant. A few male and female sparrows along with other birds were busy looking for feed. Today, nothing showed that something had happened here yesterday. I am unable to identify the male and female from other birds but I can easily identify male and female sparrows. Male sparrows have a black beard below the beak and females don't. Otherwise, the wings of female sparrows, like beautiful women, have a different light brown color, which gives a beautiful appearance. As a man can be identified from his beard, in the same way a male sparrow can be identified from his black beard.

Yesterday fierce fighting took place between male and female sparrows. One male and one female sparrow were the victims of the ire of the other sparrows. The male sparrow was miserably isolated by the other sparrows. Sparrows were attacking him together in pairs. He was looking very defensive and scared. Other sparrows made a fierce attack on him with beaks and feet and scared him to death. He was looking very weak under their attacks. He was flying from one place to another to save his life but could not get relief. A female sparrow was also facing similar circumstances and facing similar attacks. She entered under the thick plants of flowers for some relief but the other sparrows also went there to attack her. She was able to get some relief from the thickness because the other sparrows were not able to attack like they had in the open place but not for a long time. The pecking beaks of the other sparrows compelled her to come out. It was difficult for her to save her life.

Now I was wondering, what could be the reasons for their fight? Not going very deep, I guessed there might be a dispute between them related to character. It could be that their relationship was unacceptable to the others. They must have broken an unacceptable law for which they were getting this punishment. There may be another reason such as they have not cared about their fledging or they have not fulfilled their other responsibility and even did some negligent deed. That the two were defensive shows they must have done some unacceptable deed. Maybe their personal enmity was the reason for their fight. Maybe they are weak and have come under the pressure of the strong sparrows. Maybe they have come from another place and become the victim because they are strangers. Maybe all my guesses are wrong. Man can be less humiliated and disgraced by others than he is humiliated and disgraced by self-conscience. A wrong thing twitches the conscience. If there is a right way, then there is a duty also to perform it. Everybody will fall in his own grave. Man came on earth and society also took birth. Society created rules and laws to simplify human life. Society put a goad on volition and obduracy. The power of society gave negative results too instead of just a goad on volition and obduracy.

In society, some people acquired too much power in their hands and ignored the rights of others. Social comparison took place. Two things became apparent, need and disadvantage. Need became the winner. Society is the need with disadvantage. I don't know whether there is any disadvantage in the society of sparrows or not. Whether there is any big effect of gathered power into the hands of society is a disadvantage either, or their fierce fight is limited only to solving the problem.

The countries took birth from man's social comparison, which I have never seen in the society of birds. Maybe I am ignorant in this matter. Social power was exploited by some men and it was converted into countries. The system of countries captured the power of society. Now the system was looking like the society. Countries make laws and rules like society, which became the system. Many individuals took the benefit of the system and they limited the society in the shape of kingdoms and dictatorships. Greed and power went out to capture more land and wealth from other countries. Greed caused men to capture the natural resources and brought more dreadful faces in front. Now nobody cares how society became the victim of greed. Power has shown his powerful system and snubbed society. Now nature does not serve all human beings. It has become the slave of the powerful system and owners who were in possession of natural sources and of things made from it. Man, human beings and society became the grass cutters in the power's hand. They must have become grass cutters but not dead. How they can die; the king needs the public to rule. The system changed its color like a chameleon. To make the system good looking and more attractive, both appearance and arguments were present. Greed covered everything with explosives. A large mass of men and human beings exchanged beautiful arguments and philosophical views. The wounded society cried too much but who cared?

It was not the case that real philosophy did not come in defense. Suggestions came forward to make the circle of society bigger and make the shape of system like a society. To go forward, more suggestion came to finish the form family and make the society without love, affection, or attachment. To remove the weakness of sex and wealth, they tried to treat everybody equally. This way of thinking was looking very impressive. To admit the whole earth as a joint population and to give everyone equal rights impressed everybody. But nobody judged the importance of society less in the system. In a vast system in which family affection was finished, it could not be successful to convert the society into a system because the root of the system was materialistic instead of depending on nature. The foundation of society always remains in nature. Materialism in its dangerous face became comparative with naturalism but drowned in its greedy weakness.

It looks like sparrows don't need such a big system to solve their problems. I became busy in my work. I could not see more of the fight of the sparrows. Otherwise, I don't know how greatly the system of sparrows is developed or how they properly solve these problems. Otherwise, I saw another race of sparrows in which the unity gives an impression of a big society that is not limited only to one country but at the world level, in spite of the fact they are not bound or recognize the boundaries of countries.

Kingdoms and dictatorships that developed from the society and became the crisis for man brought another new ideology and developed democracy. Power was gathered into systems. In spite of the big progress and development, kings were created everywhere in which man was bound to elect those who would suck his blood. One sucks the blood and another waits a few years for his turn in a very attractive and cultured way because it was unbelievably strong system so nobody could save themselves. The rulers of these systems make sure that nobody could speak against them because they have got it stamped from the people. There is no chance now that any philosophy will take birth because this system has been adopted by all the world powers. Wherever common man lives in this world, his condition is the same everywhere. Democracy became the show of richness of a few and people have the right to elect one from them and people's voice became locked in it. Now nobody has the right to speak against it. The human being is buried under needs and good and bad things and ammunition and explosives.

Whether there is society, sociality, or system among sparrows or not is impossible to know, but there was no tension or strain in them like man, which is given to him by his thinking. I have not seen these passions even in their fighting. I have not seen that they are fitted in the society like a part of a machine, so as when one part becomes defective, it can be changed with other.

When the ideology of socialism was spreading, the ideology of the open market competed with it. The motive was to provide high volumes of trade through open markets. This means that man has no right to live anywhere in the world but he has to bear the burden of open trade. This was the result of the thought that attacked other countries for the expansion of their trade. Now slavery has been converted into economic slavery. The meaning of this is the traders of developed, undeveloped, or developing countries are able to loot using both hands but common people are forced to bear the hooliganism of those in the boundaries like the bird in a cage.

It looks like sparrows are free from the disputes of theists and atheists. Sparrows depend only on nature. Among humans, when theism could not save humanity even by sacrificing people's lives, and atrocity captured it, atheism took birth. The idea behind it was only to save humanism. Philosophy was the one behind theism and atheism to save humanity. Both competed with one power, which was that of trade's power to collect wealth.

There is one page of history on which you can enter the names of those who served humanity by sacrificing their lives by flushing their blood and another page is of those selfish people who stain that page. Philosophy always saw the humanity in the lap of nature. It never admits the life above nature.

Today everything was looking peaceful. There was no clue of yesterday's fight. The sparrows were busy in their routine life. How they solve their dispute, I have no idea. What was the result of their dispute, I have no knowledge about. How they created a peaceful atmosphere, I don't know. But today all was looking normal. I don't even know who was aggressive and who was victim yesterday. I cannot say whether those sparrows today are the same ones or not. I am totally ignorant in this matter today. But I don't think they are not the same ones. Where could they go?

A New Age

I was hearing very strange cries and the dreadful voices of human being, animals, and birds. All around there was a hue and cry. Nobody was able to understand what, how, and why it was happening. Nobody was able to understand the consequences of this happening. Everything was being broken and ruined in this occurrence and there was no way to communicate with each other. The earth was swinging like a hammock. The sky was thundering all the time without any break. All over, water was going this side to that side and from that side to this side. Nothing was in control of anybody. People were running hither and thither in panic to save their lives. They were going sometimes to high places or other times looking for other safe places like trees. All was going in vain. Water puts them sometimes in a deep place or on other time on a high place. Water was digging big trees out and these were flowing in water and looking like small branches of trees. Nobody was able to save himself from this unexpected occurrence and people were dying untimely and unexpectedly. Houses and every other thing were disappearing from the earth.

Mountains were collapsing. Ears were becoming deaf and tongues were becoming speechless and dumb.

Nobody exactly knows how much time it took to happen and on which part of the earth it happened and on which part it did not. Nothing was calm or peaceful. I was also drowning and I was unable to take a breath. I thought everything had been finished.

What I saw then was that I was coming out naked with heavy breathing from a sand dune. I don't know how long hairs began to grow on my body. I was looking like an ugly animal. I was not looking like a social animal any more. My shape was totally changed. I was very hungry but there was nothing to eat. Doomsday was over and everything was stable now. However, there was nothing on the earth. I don't know where the vegetation, flora, plants, trees, and herbage had gone from the earth. There was no sign of anything verdant. There was no sign of lushness or luxuriance, no greenery on the earth. I was dying of hunger, wandering hither and thither in search of food. However, there was nothing all around to eat. I saw some insects, ants, large ants, maggots, bugs, rats, snakes, pismires. I was bound to make them my food even raw. One day I saw a dog but I think he doesn't believe in man anymore and ran away from me. Sometimes I saw some other animals, some known or some unknown. However, no one trusts me and all were running away. I have started eating every live thing I could find and whatever comes into my grip.

After crossing a wide desert, I saw a plain area. I saw a small river too. I tried but could not find any fish or crabs in the water. However, I had seen nothing growing on the earth. There were no trees or anything anywhere. Sometimes I ate soil to stop my hunger. However, water has given me new life and great relief. Near the water, there were many live things, which became my food. It has become a little bit easier to keep myself alive. Then I saw some mountains, which were made of sand and stone. There was also nothing green growing on them. I became so weak from hunger that I was even unable to walk.

Now my condition was so bad that I started eating what I found, like insects or anything. I was not able to fill my stomach. Some animals tried to attack me. Some insects were very dangerous for me. Their bite sometimes gives me severe pain. Anybody was the victim of anybody else who wanted to survive. Each surviving creature could chase or hunt any other or hunt. There was nobody there to save anybody. Animals preyed on each other. They tried to prey on me and I tried to prey on them. Some birds also have become very aggressive. They also tried to attack me. They were attacking other animals and other birds.

There was no boundary of my happiness when I saw another man in the same condition as me. Very soon, we became the power for each other. Now we mentally became very strong. Sometimes we were feeling that we would conquer the world now. We took a very big power from each other. We both were dependant on small reptiles and other live things, whatever we could find. To bake what we caught, we used our previous knowledge and created fire by rubbing the stones. We were unable to find wood to keep it burning continuously. Our life was not better than the other animals. If we found any piece of wood, we kept that with us all the time. Pieces of wood were becoming our property. We were confident that with these we would be able to continue the fire. But that became our compulsion to stay at one place. We started living at the bank of river.

Days and months passed away. We were near our death beds. During this period due to rain some verdure, vegetation, and flora grew up on the earth. That was a golden boon for us. Now we were very hopeful. Life was becoming easier.

Now we became very happy by remembering our past days. One day we killed a snake. We cut off his head and tail and were ready to eat the rest of his body. In the meantime, however, a hawk came from the sky and took away that piece. That was a very strong hawk. I don't know how we were becoming so weak and that hawk was becoming so strong. "This will become a dinosaur one day," my friend remarked about this. "One day we will become monkeys with the hunger," he said one day. Our past days were becoming our history and stories.

One day we saw a woman. We tried to talk to her but she ran away. We were also running towards her. We tried too hard but could not succeed. Otherwise, this gave us a different type of satisfaction and hope. Now we started go away from that site in search of other human beings. Suddenly we saw a lion. It looked very difficult to us to save ourselves from the lion.

"We should go now," my wife said.

"Oh! My god, it is good I have woken up from that bad dream," I said.

My body was wet with sweat. "Are you OK?" my son asked.

"I am fine," I said. We have come to enjoy our weekend at the seashore. When my family members were enjoying the beach, I was sleeping on a piece of cloth on the sand.

Awakening

Bhikhi Raj was looking very happy today. Before this, whenever Bhikhi Raj visited heaven to accompany Rikhi Raj, he came with a serious matter to discuss. He also becomes very serious to get the answer from Rikhi Raj. But today in contrast to that, he was in a light mood and started praising Rikhi Raj by entering heaven without any reason.

Rikhi Raj gave a smile first and then asked the reason for his happiness.

Bhikhi Raj as usual without taking this side or that side of the matter gave a direct reply, "The secret of my happiness is awakening."

Rikhi Raj asked, "Which awakening has come in you that is giving so much happiness?"

Bhikhi Raj said, "Awakening has not come in me but it has come in the man living on earth. He has fixed the dates of doomsday and is telling the other people how destruction will come during doomsday. It shows how much man has awakened and how much serious he is about addressing these problems. He looks frightened and alarmed too. I think to be frightened is correct for man but I am surprised to see the progress and awakening of man."

Bhikhi Raj put another question to Rikhi Raj, "You condemn the knowledge of man and you also say that the knowledge of man is his enemy and say that knowledge of man has created the danger to the existence of earth and it seems that doomsday is very near and is likely to come very soon. But now it seems that man is too advanced from your expectations and he has fixed the dates of doomsday and now he is waiting the day of doom."

Bhikhi Raj was not stopping from telling his point of view and said further, "I agree with you that man could not be able to capture heaven in his life but the bulk of deaths of humans during doomsday and the souls of them together may create a danger to heaven and can overcome it. You will be unable to stop them then and your throne will be in danger."

Bhikhi Raj laughed loudly and Rikhi Raj was also laughing with him as both were of the view that this would never happen.

Bhikhi Raj was now looking towards Rikhi Raj with expectation that Rikhi Raj would comment further on his views.

After laughing, Rikhi Raj became serious. He generally feels happy to talk and remove the confusion of Bhikhi Raj according to his knowledge. He feels it is his duty to give the right shape to the incomplete thoughts of Bhikhi Raj.

Bhikhi Raj also hopes this is so.

After some silence, Rikhi Raj said, "Man is right and his awakening is also right. Man knows about his misdeeds very well. Otherwise, not all men are of the same nature. Right persons put forward their right thinking openly before all human beings, admonishing them to take care. But it is not necessary that right thinking will be accepted by all men, traders, and rulers. They always try to capture the right thought and try to put forward ideas with their own stamp on it to exploit and to show the greatness of their discoveries."

After a little bit of silence, Rikhi Raj again said, "The same thing is happening now. Someone must have shown the perspective picture but the trader must have used that to earn something by frightening the world. Who knows who told the reality but everybody knows who will cash in on the situation. Man has no power to understand the movement method of the earth. Man is unable to understand and reach up to

241

the working system of the earth. Man is too small to understand these matters. Earth always charges itself in many ways, sometimes by volcanoes, sometimes by tsunamis and earthquakes. Nature also gives many warnings by way of typhoons. Man has no control over these movements of the earth even if he estimates a little bit prior to these happenings. Man is inadequate in this matter to understand the functioning of the earth. So those people who have fixed the dates of doomsday or destruction of the world must have stolen the knowledge from some lover of nature whom nature gives some indications and they must have used this knowledge according to their habit to earn money by frightening the people."

Rikhi Raj further said, "Earth has charged itself recently for many years and it will save its naturalness till the end. To destroy the knowledge of man will be its last step and it will give many warnings before that and during the last warning, it will finish all. Nature will not allow man to interfere into any of its functioning. There is no meaning of man's estimates on earth. Nature is independent and man is mortal and confined into its knowledge. Nature is able to maintain itself without the help of man."

"You are saying that there is no immediate danger to the earth for several hundred years and if in several hundred years doomsday will not come, and then man will automatically be finished by being frightened by such a long wait," Bhikhi Raj said and they laughed boisterously.

"He will not be frightened by himself but natural calamities and disasters will teach him the real lesson and will teach him how to behave with the nature," Rikhi Raj said and they started laughing again.

"To show that man has become too advanced and to show his victorious behavior, sometimes he bragged like this. Man shows that he is too forward from nature and is able to control nature but in reality there is no endurance in his professed accomplishments because the pillar of knowledge on which he is standing may be demolished in a second by nature," Rikhi Raj said.

Bhikhi Raj said, "Man is professing on the basis of his knowledge that he is able to control all the movements and is able to cure problems like global warming, thinking that the earth and nature are like their machines, which they are able to repair. They are arguing like this as if everything is in their hands when the reality is there is nothing in their hands."

"Now you have become wiser than me," Rikhi Raj said.

"Excuse me! Excuse me! Kidding is not allowed," Bhikhi Raj said and both started laughing.

Man Knows Man

I am not frightened with the action on my roots. This is the only way to reach the destination for me. I have learned this through meditation. We always arrive at the destination by doing meditation. The way to complete the meditation is only to serve with confidence without selfishness. No one can do this easily. We also all bear the same calamity that life suffers. Actually, we don't believe any calamity to be a calamity; rather, we treat that as a routine of life in which there is hidden goodness and welfare for someone. The service will not be finished with the end of my life. Through my death, all nature will be served. Maybe a big part of me will become the part of the material of life but my services will remain to observe the rule of nature.

I grew up in the jungle. It is not necessary that every tree has to grow in the jungle. A tree can grow anywhere. Actually, trees are grown everywhere to serve nature and its creation. A wood cutter is very happy to see my big size because he will get a huge and good quantity and quality of wood. All possible buyers are convinced easily from the wood like mine because it is very expensive on the market. Otherwise, nobody can give the price of a tree. Nobody can give the price of his breath and I am the creator of his breath. It is the needs of man that compelled him to fix the price of a tree.

I don't care about the attack on my roots. The philanthropy that I did in my life, I will do more than that after my cutting. Both my life and death are a form of philanthropy for man. Man during life gets everything but is not able to give anything and after death, his ash is unable to do any service to anything. Man is afraid to see the graves of man. His collections and possession upon earth give him no support after death. The needs of men have divided them up to the extreme level that they have started killing each other out of their selfish motives. Trees are not divided like men to fulfill their needs. A needy person cannot fix our price to that of someone whose work is only to give service to others. A man that is becoming happy because he will get a huge price from my wood is a selfish man and he cannot stop my selfless services even after cutting and when I will be lifeless. I don't mind if I will be the main door of a king or the throne, a stick of a hut, a stick of a nest or whether an animal's cud will be chewing my leaves. All these things give me equal satisfaction and no one can put any hurdle in my way to do service. In all these ways, there may be selfishness of someone but there is no selfishness in me or in my service. I am not in the ownership of anyone but someone needy or selfish may claim his right to me. From using an axe on my roots and up to taking the price of my wood according to his desire, he cannot change the way of my selfless services. I don't care how I am used but I am happy someone is using me and taking my services.

Maybe the secret of happiness of the man who is using his axe on my roots is one of my qualities by which he will be benefitted but we are not bound by our qualities. Which tree has which quality is not our objective? This quality determines who uses our service, as well as how, when, and where he will accept our service. Our size or quality may be our identification but our services are selfless for all. Maybe vanity is the fruit of the needs and selfishness of man, which reflect the good and bad side of his society, but our selfless services are the part of nature that is above the perfection of quality or size.

The man who is cutting my roots with an axe might not know that man had a dream and aspired to solve the world's problems by learning from our meditation. Maybe this relationship is the secret story of our good relations. We have no relation with needs and selfishness but man has made these his motive. The man that becomes like us due to our relationship sees all virtuous deeds from solid fuel up to the fuel of a funeral pile, which is the motive of our life.

If the man who is cutting my roots understands the meaning of giving the services instead of getting by grabbing, then he might be able to understand the meaning of saints and sages. If he is unable to understand the desire of human beings and depends only upon the materialistic life, then what could he understand of my life's meaning? However, there are saints and sages who could not find selfishness in us and so made their life like us.

The man who is cutting my roots and will be happy with the good price may not know that our services are not limited to man. Our services are beyond the limitations of man. Maybe he doesn't care about the creation. Maybe he is unable to think beyond his needs. My services do not have a double standard like that of man. The man who is cutting my roots may not have a double standard but he does have a short-term way of thinking. That is why I have sympathy with him and I have no doubt that it is because of his short-term thinking that he is cutting me only to get some money.

My work is only to serve anyone and if he will earn something for his livelihood by selling my wood after cutting me, that will also be a good thing for me. But the man of a double standard watches the world only with a selfish mind and if he claims he is sympathetic to the creation, he is a big deceiver and hypocrite. He deceives not only himself but the whole creation. There is only one difference between a man and a tree, which is that a tree has no cleverness like man and man always suffers for his cleverness. The benefit of cleverness comes to very few but the whole of creation suffers. I know the man who is cutting my roots doesn't know my importance to the world.

Trees don't need to be theists or atheists like man to fulfill their needs. Their services are for everything. These are the puzzles of man and they are busy to solve it. If somebody cuts trees, even then it can be considered fortunate because this action will come in use for someone. A tree can pass centuries at one place. Man treat it as meditation that a tree does so many services standing at one place and it demands nothing in return for these services.

I don't accuse the man who is cutting my roots or human beings in general. I don't want to put man in the position of being a defendant versus creation. Actually, all of creation depends upon all of its parts and nature supports this cause. It looks like everyone in the creation living his life and supporting each other to live in some way. Out of creation, only man has such thinking that can disturb the whole world and has polluted every source of nature. All sources of nature are afraid about the destination of man. The earth, which has enough resources to finish the pollution, is feeling unable to complete its mission.

The man who is cutting my roots doesn't know about these dangerous warnings and it looks like this action concerns his livelihood only. I am also not able to understand the deepness of these matters. I felt impelled to say something during the time of cutting, although I cannot put into question the man who is cutting my roots. A big part of the human beings in the world is still unable to get the fruit of natural resources due to being in possession of so few, so how I can put all human beings into question as they are struggling? When man is not allowing the unending resources of nature to reach up to man, then how I can blame all? I cannot forget that man also helps me in growing up. However, maybe man's wrongdoings are heavier than his good doing, which buries all that.

Now I am near to falling. The man who is cutting my roots and taking short rests is very near to success. I will fall soon. I don't mind if I will become a chair or table or become coals by burning in the fire for man. He can burn his furnace or funeral pile with my wood. I will come in use in all manner of ways. He cannot live without me.

I am the shareholder of the progress of man but maybe man's doing will make it difficult for him to take breath. Nature is unending and will remain unending but man can become his own enemy. Man knows man. I am now falling and the time has come of my rest.

Dwarf Man

Bhikhi Raj, in a furious mood, went to Rikhi Raj and said, "Man has enough resources to eat, live, and enjoy the life over the earth but why does he all the time start war in the sky with other planets?"

Rikhi Raj, understanding the trend of his talk, said in the first instance to calm Bhikhi Raj, "Now you are thinking in a very advanced way. There is no need to tell you anything more. You know everything very well."

Hearing Rikhi Raj's tone, Bhikhi Raj really became calm and said, "Man thinks himself very knowledgeable but all the time he commits the same wrongs that change the fate of earth and he destroys himself."

Noting his friend's changed mood, Rikhi Raj and, wanting to know more about the topic in his mind, said, "What do you mean by all the time?"

Bhikhi Raj understood that Rikhi Raj had caught his right pulse, so said defensively, "A sky war is going to be fought, but not for the first time. This has been fought so many times before this. Up to now, it has been the deities and goddesses that took part in it. The truth is that the sky has power to cut the wings of man and to compel man to remain on the level of the earth. Man's dreams to rule the sky all the time was crushed by nature. The reality is written in nature and kept on earth to remind all of the wrongs of man and to prohibit him from committing the same wrongs in future. But what has man learned from that? He is running the same race and is trying to capture the sky to rule over there as he wishes that he will actually be successful this time to win the sky. How foolish he is!"

"Why are you worrying about that? He will bear the consequences of all these actions if he will do anything wrong," Rikhi Raj said to search his friend's mind more.

"I am not worrying about myself. I think, your throne is in danger," Bhikhi Raj taunted back.

Both laughed together.

"You have become very clever," Rikhi Raj cajoled Bhikhi Raj.

"Pardon me, sir," Rikhi Raj said.

Rikhi Raj became very serious and said, "Your comments are totally right. The sky is invincible and man is an idiot. These both are eternal truths. The sky cannot be captured and on the other hand man cannot get rid of egoism. Greed, egoism, anger, lust, attachment are the enemies of man. He cannot get away from these evils. Almost all evils are powerfully controlling man. Even if he got rid of some of these, others are enough to finish him. Even one evil is sufficient to finish man. When man stands on the heap of knowledge, then even if he got rid of all these evils, he becomes destitute under the burden of knowledge. Man's personal knowledge becomes dwarfed before the gathered heap of knowledge. Man cannot save himself from the consequences of gathered knowledge. This is the big tragedy of man. Truth also fights with the stick of the same knowledge with lie. So truth also gets no good result. The damage done by the gathered knowledge finishes all. Truth and lie go in the same grave as the mouth of truth and lie was towards the sky instead of the earth. Whenever man tried to leave the earth to capture the sky, the sky never tolerated it and compelled him to live within his own limits. The sky always broke the knowledge of man even if the earth had to pay the price of it. This has happened again and again and I don't know how many times more it will happen and what will be the consequences."

"Is the sky is so dangerous for man?" Bhikhi Raj asked.

"Not at all, the sky always helps every life upon the earth to live. The earth rotates around its central pole and revolves around the sun in the sky. But the earth does all this under the rules of nature and it is not wandering hither and thither in the sky without following the traditional routes. The earth doesn't break the natural rule and it never allows anything else to interfere with its routine. So it is enough for you to know that earth and sky will not allow man to interfere with their disciplined routines. All of nature is helping the earth to maintain all its routes in the sky. Every life upon the earth is part of the earth. Every life takes birth from the earth and after death becomes earth. The life that is part of the earth will not be allowed to create danger on the basis of gathered knowledge to its own creation. This has happened on the earth so many times. Man tried to fulfill his wishes but failed every time. Now again he is trying to do the same and he will never succeed in his evil designs. Man is digging his grave and even for all of creation, due to his greed," Rikhi Raj said.

"Man is trying to reach up to the stars on the basis of his gathered knowledge," Bhikhi Raj said.

"This is his big blunder. He has not such power to go out from his planet. This blunder can finish the beauty of the earth. Everything can be finished in one moment. Man is mortal. His knowledge is unable to save him. His knowledge has no value in the eyes of the natural process," Rikhi Raj said.

"Then man has to learn to live on the earth and keep his eyes down. But I don't think it is possible," Bhikhi Raj said, in a taunting way.

Both laughed at the same time.

"If he looked towards sky, the violent sneeze is a must," Rikhi Raj said in the same way and both started laughing boisterously.

"Poor dwarf man," Bhikhi Raj said.

Pulse of Time

Bhikhi Raj was very happy with his tour outside heaven with the company of time and was unable to keep himself in clothes He was very eager to share his knowledge and happiness with Rikhi Raj. He went to meet Rikhi Raj immediately.

As he had hoped, Rikhi Raj was there. He also had come back to heaven.

"Your honor is looking very happy, what is the matter?" Rikhi Raj asked Bhikhi Raj.

Bhikhi Raj was unable to find the words to start his talk in a beautiful and effective way. He said, "I have come from the tour of the earth with the company of time and I thought I should share the experience with you."

"It is very good. You should share the experience with me. Actually, I also went to see the pulse of time. Time was very disgusted these days. It was necessary to see it," Rikhi Raj said.

Bhikhi Raj's zeal was fused even further by hearing these words of Rikhi Raj. When time was leading Bhikhi Raj, Rikhi Raj was there to see its pulse. He was wondering about which experience he would now tell to Rikhi Raj. He was now feeling annoyed. To continue the talks, Bhikhi Raj said, "I should then hear your view first. I went just to enjoy but you have come with great achievement."

"True, you are totally right. Truly, it was a very important matter. It was very urgent that I understand the situation. It was important to understand the disgusting situation of time in its present. In the past century, very vast changes have appeared that have had a very grave effect on the earth. In the coming century, the circumstances will be even more pitiful. So it was necessary to discuss the matter," Rikhi Raj said.

"Centuries and present—what you mean?" Bhikhi Raj asked in surprise.

"Your memory is very short, like that of man. When you are talking about time, you must not think in the short term. Its pages are of centuries. You can watch time only like this. Your thinking is like man, who always thinks keeping in view his life time and calculates nature within that timeframe. They have fixed the dates of doomsday like that to surprise the coeval. When you are calculating the work of time and see its pulse, you cannot calculate keeping it within one man's age. You can watch time keeping in view the age of time. Then you will see even centuries are a very short term to see the truth," Rikhi Raj said.

Now Bhikhi Raj's brain was totally washed out about his tour. He was waiting eagerly to hear further. He cut short his eagerness and asked, "If I were as intelligent as you, then why would I ask this question?"

"No, it's not true. You are only my inspiration. If you do not discuss with me, maybe these thoughts will never come in my mind. You are very intelligent and virtuous for me," Rikhi Raj said very lovingly.

"You told me that you went to check the pulse of time. This has increased my eagerness to know the result. Please clarify this and reduce my curiosity," Bhikhi Raj said, putting the point to a direct question.

Rikhi Raj smiled as usual and said, "I am a fan of your excellent thinking. You said the exact matter very easily. Yes, I have come to know many things. There was lot of difference between the start and the end of the century over the earth. As you know, the sky is full of other earths and suns and this earth is a small part of the sky. In the beginning of the century, there were not very many big miracles of science but in the end, apparently it came in front of nature to beat or destroy it. Many dangerous results came forward. Now it is not hidden truth or myth that was at the beginning of the century that people were thinking. Science has repeated all these things again and proved that the previous happenings were not a myth. That was also the progress of science. Now it is everlasting truth that is unavoidable."

"What is the relation of it with the pulse of time?" Bhikhi raj asked.

"It is the present of time. It has directly linked with it today. The role that the last century played and what will happen until the end of this century, no one knows. As there was nothing perceptible at the start of the last century, in the same way, no one knows what will happen till the end of this century. This is the reason for the agony of time," Rikhi Raj said.

"I could not find any wonderful thing from it. There is nothing new. You have already told me this many times," Bhikhi Raj said.

"You are speaking very correctly. Actually, the matter is not limited to this. The real story is different. We cannot know everything from the pulse. You have to collect more information from the patient. What time told me is that there are many tomes and sacred books lying on one part of the earth in which there is written very important information about all of creation and nature. Man could not discover more information till now from that information. Man could get very little information from those tomes and sacred books. People of other parts of the world, although, want to cross that limit but they will not succeed at all," Rikhi Raj said.

"If they are not able to succeed, then the matter is finished. It means there is no danger," Bhikhi Raj said in a sarcastic way.

"It's your weakness that you get the result before hearing the full matter. You should make up your final mind after hearing the full truth," Rikhi Raj rebuked.

"Excuse me, sir. Tell me more," Bhikhi Raj apologized.

"People from outside that area will not succeed but I have not said that people of that area will also not succeed," Rikhi Raj said.

"What is the difference of people of that area and people of another area?" Bhikhi Raj asked.

"There is a big difference. The major difference is of real and artificial life. Outside from that area, life has become dependent on contaminated food but people of that area still depend on natural food prepared fresh daily from natural resources. Readymade food delinks man from nature. People of that area are still directly linked with nature. Man's waste is the food of plants and plant's produce and air is the life of man. Both are directly linked. Neither needs the middleman. This is the rule of nature," Rikhi Raj said.

"What is the effect of this? I am unable to understand," Bhikhi Raj said.

"Yes, you will not be able to understand yet because I have not told you that which will be the real threat," Rikhi Raj said.

"Tell me about it and bring me into the light, please," Bhikhi Raj said.

"Thinking linked with nature remains intense and intellectual if it remains in the natural course. But if it does not remain linked with nature, greed gives birth to demons. Demonic thinking is the real threat. There is an end of ego but there is no end of greed. Greed is able to finish the whole world. As the people have that kind of intense and intellect thinking, so they are able to discover other planets in the sky. Their action will not be in response to scarcity but will be due to greed. Their greed may finish all, which was the anxiety of the time," Rikhi Raj said.

"I am still not able to understand anything," Bhikhi Raj said.

"You must understand. Nothing can go or come in or out from the planet, as it is an infringement of the law of nature. But only their intellect and intense thinking is able to cross this line. There are many valuable things outside our planet to convince us about that. Their greed may compel them to cross the limit due to their selfishness, which may damage the balance of this planet. This I could see from the pulse of time," Rikhi Raj said.

"I am unable to hear more," scared Bhikhi Raj said and left the site.

Tour

Today when Bhikhi Raj visited heaven, he could not find Rikhi Raj. This was not the first time this had happened. Many times before this, Rikhi Raj remained absent for many days without telling Bhikhi Raj anything. It was their habit not to ask anything about these absences.

An idea stroked the mind of Bhikhi Raj. He decided to go out from heaven to see and enjoy the beauty of the earth. On the main door of heaven, when he was leaving, the time welcomed him to lead him. Bhikhi Raj accepted the guidance of time happily and offered thanks for the services of time.

Bhikhi Raj saw the globe of the earth rotating around its axis and revolving around the sun making days, nights, and seasons. It was looking very beautiful. The moon was also revolving around the earth. There were many black holes in the sky, which were helping the moon and earth to find the correct path and to remain on the correct line and to keep the exact distance without any interruption. Bhikhi Raj was enjoying the miracles of nature and beauty of cosmos with the company of time.

On the way, time told many stories of the earth from the beginning to the end to Bhikhi Raj. Time told Bhikhi Raj that everything on the earth is the product of the earth and after enjoying life it again mixes in the earth and becomes part of the earth. What he will mix in the earth is the destiny of that life, which decides his deeds over the earth.

Bhikhi Raj asked time, "Do you know why theists and atheists take birth on the earth?"

Time smiled over Bhikhi Raj and said, "This question is related to man. In the circle of nature, all are equal. In the eyes of nature, there is no difference between anyone. From birth to death, every life plays the part of a role. During that play, theists and atheists take birth. I, means time plays his part of role. Theists and atheists help life to decide their path and become busy with their activities. Earth has no discrimination for anyone. Sins and virtuous deeds are the personal earning of every life, which goes into each person's account. Some souls get a good life due to virtuous deeds and some get a bad one due to sins. Every soul has to bear the heaven or hell of life.

Bhikhi Raj heard a hue and cry, "Ate, Ate!" After asking time, Bhikhi Raj came to know that "ate" means that which he does not understand. In the mind of Bhikhi Raj, the meaning of ate was only about eatable things. He was thinking earth had produced enough things to eat. How could anyone eat more than his capacity? He thought it would be better to clarify this more from time.

Time told him, "It would be better for you not to be confused in these matters. This fight is not about eating. How much could one eat? This is a paper fight. To control the natural resources of earth, man has developed paper currency and this fight is about that currency. Most of the fights are due to these paper currencies. Man treats this currency as a second god."

Bhikhi Raj was very surprised to hear this. He was of the view that the earth has produced everything to nourish the life on earth and every life depends on each other. He was of the view that there was only a crisis when one makes another his food. He was also aware about the possession of man over earth but he heard for the first time the fight over paper.

He saw over the earth thoroughly. His eyes dazzled with the light of electricity over the earth. He remembered Rikhi Raj's view. Rikhi Raj called this the indigence of man. How many efforts he made to get this light but he could not compete with the sun of nature. He never thanks the sun and nature for this marvelous blessing. With one glimpse in the morning, the sun gave such a huge light that the back side of the earth became dark. Man's act was a supplication before the sun.

Bhikhi Raj saw the earth more deeply. He felt the science of man as toys to play with. Man was keeping himself busy with these toys. He was running with, on and after these toys. He felt that man is so busy with these toys that he forgets himself. He felt sympathy with lost man who is living an artificial life.

Bhikhi Raj remembered Rikhi Raj's view that man has squeezed everything from earth and assembled absurd toys and brought extreme danger to the earth. Every act of man is the bell of danger to nature. Earth is feeling uneasy to rotate around its central pole and revolve around the sun. All the time, there is major change in the seasons and weather. There a lot of black holes in the sky that help the earth to keep the correct path but due to this uneasiness, at any time, these can be penetrated, which can give major harm to the earth. Water, which is playing the role of father for every life, and air, which is the breath of every life, and the sun, which keeps life running and all jointly keeping the earth moving and maintaining the balance in the sky may all be damned by the actions of man, which make all of them incapable of helping, and so the earth fell in great danger.

It was the truth that the earth liked this status very well but the weakness given to it by the damned acts of man and pollution may change its character and cause it to fail to maintain its balance forever in spite of the fact that the earth loves its creation very much.

Bhikhi Raj saw that there was a great deal of unrest in the minds of all other aspects of creation due to the acts of man, as man had already finished many species upon the earth forever but all were helpless to do anything and were bearing the cruelty of man.

Bhikhi Raj became very sad by thinking all this and decided to go back. Time said good bye at the main gate of the heaven to Bhikhi Raj.

Toga of Words

Sitting on the sand at the shore of the sea, I was looking at the waves coming towards me. The water comes forward with full force and goes back within the shore in a few moments. My mind was also buoyant like the waves of the sea. The thoughts coming in my mind were also not measurable like the water of the sea. I don't know whether the earth is rotating and revolving in the sky with the help of water or water is rotating and revolving with the help of the earth. In this rotating and revolving, I don't know who helps whom. But whenever anyone talks about this, they mention the name of the earth only. Nobody ever said that water or sea is rotating and revolving in the sky in spite of the fact that the quantity of water is more than the earth. Man always prefers the earth as he lives on the earth and uses the name of the earth accordingly when talking about his. This is not actually the issue at this time in my mind, whether the name of the sea should be there too. I only want to see and enjoy the present moments of wonders of nature.

If you actually asked me at this stage when my mind was so indulged in the charming scenes, I think I don't need any toga of words to describe the beauty of these moments. If I would use any word to explain this, those would only show the weakness of my mind. My joy incarnate cannot be disclosed by words. There was no importance of words for me at this time. I was only in need of mind to enjoy the beauty of the present moments. Eyes and ears were needed but a tongue was not necessary to enjoy the moment. The scene was giving me peace of mind, which cannot be tied with words. In front of me there was water and water, at the backside there was land and land and in between on the shore where I was sitting.

Waves coming towards shore were creating a wonderful musical sound, which was providing me more peace of mind. Whenever anyone would have researched the musical rhythm, then he must have taken the melody from these types of miracle of nature. Actually, real music has no need of any instruments. You can find the music near falls and rivers. You can hear the music in the sounds of birds. You can find the musical rhythm in the movements of deer, horses, and camels. You can hear the music of the air passing through the trees. Nature has spread music everywhere. The real truth is that music made by instruments cannot be more beautiful than the natural music.

I have read many stories of saints and sages who meditated so much that during long meditation, they were buried under soil and grass grew over them. They meditated in such a peaceful and calm way that they forgot everything around them. I am not so calm and devoted like those saints and sages and may be unable to become so because I am too materialistic and cannot be so devoted like them. My front side is towards the sea. In front of me, there is water and water and on my backside, there is land and land. I am very calm and indulged in the beautiful moments. Actually, everyone gives importance to his front side so everyone thinks all happens in front of man. Eyes to see are on the front side of man. Mouth, nose, ears, forehead, stomach are in front side of man. Arms also work towards the front side. Legs and feet take steps towards the front side. Is that why I said that everything happens in front of man? Mostly, on the side where all body parts work, we call that the front side and the other side is the backside. I am feeling that my mind is enjoying the situation that is happening in my front side but I am not lost in it to such a level that I have become oblivious.

My front side at this time needs no words to explain the moments happening before me. I want to enjoy sitting at the harmonious meeting shore of earth and sea and don't want to bring anything in between. I want to enjoy the situation with my mind without the interference of words.

My backside is also part of me. I cannot think anything by separating my backside. I am telling this because the condition of my backside is not the same as of my front side. There is a recitation drilled for me to learn by rote at the backside of my body that I cannot leave apart. I cannot ignore that recitation that has been drilled into me. I cannot even ignore the meaning of that recitation drill. My front side only is looking calm at this time. I may say that I am ignorant, illiterate, or uncivilized, even foolish, but it is difficult for me to ignore the drilled recitation. I also cannot say that I was not told the meaning of the recitation at the time of drilling. I can forget the recitation drilled for the time being, I can ignore the pain of it but I cannot separate that from myself.

I am sitting on the shore, which can become only at the meeting point of sea and earth. I am sitting on the earth. I cannot walk on water. My body is such that I cannot walk without the earth. I can swim or can go by other means in the sea but the earth is my last destination. Fishes can say that the sea is rotating and revolving in the sky and not the earth. Their last destination is water and my destination is earth but there are many common factors in us too. Air is common, which is present everywhere without any discrimination. Sunshine is common, which spreads all over for everyone. Light and darkness are common, sky is common. Otherwise, there is a deep relationship of earth and sea. Neither can be accessed separately. It looks like all these are useless ideas of my mind. Otherwise, a fish will never be satisfied without licking the stone.

What is the relationship of the sea and me? How many people die without seeing the sea in their life? My backside is towards the earth where I have to return but my front side is towards the sea. Maybe it is not possible but I am trying to ignore the recitation drilled into me and the words to come in the way at the present moments of enjoyment.

The recitation drilled on my back is related to words. The backside and the recitation on the backside have no existence without words, so my backside is covered with the toga of words. The reality is that the toga is not very clean. There are many blots of discrimination on it. Otherwise, the toga is stitched with ego. The cloth is also composed of selfishness and attachment, which is not durable. So it will be finished soon. The color of the toga is full of lust and wrath, which looks beautiful and forceful but not meaningful. I don't think that these blots can be cleaned. Even if I become able to clean some of it with severe efforts, the color of the toga will surely be absurd.

It was my own prediction that the soul in the toga always remains sacred. So I made this toga keeping in view the glimpse of some sacred togas. Somebody said this truth, that a crow cannot be a swan. I was well aware of my weakness. The soul in a toga never allows any blot on the toga. They choose to color the toga with blood first to save the toga from blotting. The color of blood never allows the toga in any age to be dirty. The sun of truth is never drowned. With the sweat of selfishness, the toga always gives off a stench. I was able to live in ordinary dress in which the sweat of my labor can be consumed and that can be cleaned by washing easily but this did not happen due to my selfishness. Maybe it was not allowed to enter my consciousness as a resentment of the drilled recitation.

My heart is not working like my thinking. It is singing and dancing before me and saying openly that I did not wear the toga only for a selfish purpose. Nobody colored their toga with blood by themselves and the time has not lapsed so far. I was still in the line of billions of people who were bearing atrocity, drowning in a marsh of power and were helpless to escape. So I wore this toga as a proclamation. But it was not that toga that has the strength to encounter with the rocks of power. Nobody had time to hear my proclamation. They were bearing the atrocity with humility by pretending their attention. My toga of selfishness was finished with their toga of patience. There was more power in their patience than in my selfishness, which mutilated my toga. My toga was torn with the stump of their ordinary dress. My lie was dispersed in their truth. My toga became a frightened truth.

In whose hand I saw the upright post of the plough and wore the toga to defend by watching his fate flying in the air, but my toga became hopeless when I saw the plough running on his body. I wore the toga by seeing that there is no price of the sweat of labor but my toga became ashamed when I saw that some kneading their flour in his blood. There may be other truths that made my toga ugly but at this time, I want redemption from the drilled recitation and toga of words.

I was not alone here to enjoy at the shore of the sea. Many birds were also enjoying these beautiful moments with me. Maybe nobody drilled a recitation on their body? They also don't need the toga of words. Maybe this is the only reason that they were enjoying more than me.

Driver of Nature

Bhikhi Raj was standing on a wall and giving a lecture to the passing people, who were not even taking notice of him. He was not paying any attention to the rifles of the police that were targeted on him. He was continuously speaking without a break; "If you want to save the earth, return its things back to it that you have dug from it and use for your life of luxury. Refuse to use these things any more. These things are made from the strength of the earth, from which it took energy to do its routine work to nourish the whole creation. Things made from iron and other metals beyond limit are useless. These rob the strength of the earth and weaken it. You are using the strength of the earth for the luxurious life style whose price you have to pay and the earth has to bear. Return it to the earth. Send it back to the earth through water of the sea otherwise every creation of god will have to suffer. Blindly making and using these things will give severe consequences. You can live easily without the things made from the strength of the earth. Your life will never be affected if you will start living without these luxurious things. Your hobbies may too expensive in the near future. Your needs are not too much but you are using the earth's strength too much without any need. If you will not give respect to your mother, you will suffer greatly. Why do I say only you? Actually, all of creation will suffer. What are your needs? Go and think. How much do you use that you don't need at all? Go and consider this. Your needs are minor but you are wasting too much the strength of your mother. Throw all cars and other metal material in the ocean and use the minimum resources of nature. Be the driver of nature today."

He ripped his shirt from the front side. On the vest was written, "Driver of Nature." Actually, one day somebody gave a new vest to Bhikhi Raj. Rikhi Raj wrote on it, "Driver of Nature." Bhikhi Raj never wore this openly. Today, he tried to give the proof and recognition to his thoughts.

People passing by were laughing at him. The police were asking people to beware of him, and telling people that he is mentally ill and might do anything. They were saying that he had not yet become violent but they have come to put him in the hospital.

Actually, Rikhi Raj sometime ago had died as the result of excessive intoxication. When Bhikhi Raj saw him in the morning dead, he ran away in fear, leaving the dead body of Rikhi Raj in the hovel. He was aware that as a homeless person he was not able to cremate the body of Rikhi Raj. It was in his knowledge that cremation was very expensive. He was aware that the dead body has to be kept in the mortuary for some days until the time for cremation comes. He had nothing to spend as he depended on daily begging for his livelihood. Also, he was in fear that Rikhi Raj had died from excessive intoxication and the police could hold him responsible for giving that intoxication. Actually, all homeless people do the same thing as Bhikhi Raj. On the death of a homeless person, they inform the police one way or another and leave the body alone. Nobody takes any unnecessary risk of being subjected to any type of interrogation. Some charity voluntarily comes forward and performs the last ceremonies. A homeless person is not a martyr so that no one will raise questions about their cremation, whether this or that was not done.

Bhikhi Raj also ran away from that place and someone else informed the police about the death of Rikhi Raj. However, the police came to know that Bhikhi Raj was present there on that night of the death of Rikhi Raj but they had no suspicions of Bhikhi Raj so they took no action. The post-mortem report gave no indication of any suspicious circumstances.

Later on, they searched for Bhikhi Raj and asked him a few questions about the death of Rikhi Raj just to complete the paperwork.

The death of Rikhi Raj hurt Bhikhi Raj very badly. Actually, he was very much dependent on Rikhi Raj in many ways. It had become habitual for him to depend upon Rikhi Raj. He never expected the death of Rikhi Raj like this. It was an unbearable shock for him. Days and months passed but the tragedy of Bhikhi Raj was becoming more serious. His condition was very miserable. The habit of using marijuana already made him mentally very weak but now he was unable to get high regularly. He did not have as many sources to get marijuana as had Rikhi Raj. He became very weak and his body became very painful. He became mentally ill. It was an even bigger shock to him than the death of his father. At the time of his father's death in his home district, there were many people to console him. Here, though, there was no one to help him anymore. He thought his world was finished. He was in this country illegally. He felt a deep loneliness.

All these thoughts made him very mentally ill. He was already finished by becoming intoxicated too often. But now, by this shock, he lost all his mental balance.

The useless and rootless talk that he and Rikhi Raj shared, he was now speaking openly in public places and on the roads but nobody took notice of it. He was not even wise like Rikhi Raj who had been able to control the situation and could not talk or thought like him. So nobody understood his talk.

Sometimes now, he stole minor things from stores and houses and sometimes broke anything without any reason. He did many other abnormal activities that were unacceptable. Sometimes, he sexually assaulted ladies without any reason. People nearby complained about him. So the police took action and came to catch him to put him in the mental hospital. He held a metallic rod and he ran away when the police tried to catch him and he climbed on a wall and addressed the public. On both sides of the wall, policemen aimed rifles at him. One was trying to engage him in talk and to get the rod from his hand or to catch him safely. But he was addressing the people without caring about the police.

Heaven

Today, Bhikhi Raj was willing to talk about heaven.

He asked Rikhi Raj, "Why do some people think that if they will get entry into heaven, they will find fairies there? Some people think they will get ambrosia and other tasty food there."

Rikhi Raj as usual first smiled and then started to answer. "Actually, it is not heaven's problem. This problem is of people living on earth. There are many types of hunger on earth. One hunger is of the stomach and another hunger is of sex and greed is also a big hunger. People want to fulfill their wishes freely and without any interruption. There is struggle in the society to fulfill their wishes. Those who are not satisfied or could not fulfill their wishes of hunger on earth they try to presume that they will get that in heaven after death if they have to sacrifice over here. Actually, the hunger of everyone cannot be fulfilled because every morning comes with new hopes and new hunger. Maybe some people used these types of words to motivate some people and to inspire them to some specific cause or some people for their greed."

In the meantime, they heard some voices outside. The police were there. Both came out from heaven. Both smiled by looking at each other. Both were aware of what was going to happen next. It was a usual thing for them. They were both homeless. They made a hovel with plastic bags in between the bushes to stay in, especially during the night time and bad weather or to take some rest during some free time.

However, it was on private property. There was a big sign board too on the property, on which it was clearly written, "Private property – No trespassing."

They had been living here for two or three months. They did have not any intention to claim any right or title over the property but just made a temporary hovel to stay in and sleep at night and were aware of their probable fate. Usually, they made hovels near roads or on public properties that nobody cared about. But this time they made their hovel here because this property had not been used by anybody for a long time. Nobody had said anything to them until today. They always called their hovels "Heaven." When they sat in heaven, they started talking like they were actually living in heaven. Their talks always make no sense. It was the way of their life style. They were also addicted to marijuana. When they took a high dose of marijuana, they talked a great deal, on subjects that only they could understand; no one else who overheard them had any idea what they are talking about.

Bhikhi Raj was totally illiterate and was very dependent on Rikhi Raj in many matters. He was unable to speak and understand English. He was even unable to buy some common things from stores. Generally speaking, he was unable to mix with other people. Rikhi Raj, however, could easily speak and understand English. They used to beg at different places. Rikhi Raj usually begged near stores but Bhikhi Raj begged at some road crossings with a cardboard sign that read, "Hungry – Needs Help and Food – Please Donate." This was always written by Rikhi Raj on the cardboard for Bhikhi Raj. Rikhi Raj was also very helpful to Bhikhi Raj in other ways and in the same way, Bhikhi Raj helped Rikhi Raj in many ways. Each man needed the other very much.

Many years ago, they came to this developed country from their developing countries. Both were from different countries but understood the language of the other a little bit. They came to this country at different times. Both tried to work in this country. The circumstances for becoming homeless were different for each man. Bhikhi Raj was illiterate so he could not successfully find a regular job. Not only that, but he was a shirker. Many times, he was fired for small thefts. Many times, he was fired for irregularity. Ultimately, he became a permanently homeless person and started begging. Rikhi Raj was a smart guy but his luck had not been good. He failed from all sides. He had his own business but lost it. His wife left him. Also, he was very addicted to marijuana and sometimes he drank whisky or indulged in other intoxicants and ultimately he also became homeless due to his careless habits. By chance, they met each other and started living together. They both loved each other very much and could not live apart from each other. Bhikhi Raj had not been a habitual user of marijuana but in the company of Rikhi Raj, he also started taking marijuana.

In their home countries, neither of them had been a beggar nor homeless and had come to this country with high hopes but circumstances had not gone according to their hopes and they had become homeless. This was not the first time that they had been evacuated from this place. They were usually evacuated from every place after some time. The police didn't allow them to stay a long time at one place in spite of the fact that they were not criminals. They were not harmful to anybody but compelled by circumstances to live like this.

Today, the owner of the property must have complained to have them removed from his private property because the cop was saying that they must not have seen the board. Usually they never objected about the evacuation and the police also did not take any action except to evacuate the two men from the place. The police were not very strict towards them except to enforce the evacuation. Luckily, today they do not have any marijuana. Even so, the police usually do not make a personal search of the two men. The police know about their habits and don't want to put them in jail because doing so is just an unnecessary burden for the government, too.

They tried to take away the hovel materials but the police stopped them from doing so. They were now standing at some distance looking towards the action of the police. They were both smiling and looking at each other.

Bhikhi Raj asked, "Where we will go now?"

"Here was not your maternal house and it was not your ancestral property. We can go anywhere. The blue roof is open for us," Rikhi Raj said and both laughed boisterously.

After searching and completing the paperwork, the police returned some of their belongings but took away with them the materials of their hovel. Police warned them not to install a hovel or to stay on this property again.

Rikhi Raj and Bhikhi Raj left the spot happily with no grudge toward anybody in search of another heaven under the blue roof of the sky.

Behave

There is no difference now in whose boundaries and in whose law and rules I compelled to live. How many stories can I tell of human beings?

This universe and train of nature; I do not know how many travelers have taken a ride on it and disappeared. More travelers will come to take a ride and will then disappear. This train will always remain full like this with travelers and there will be some history left behind, some progress and society. Otherwise, you may say this world and universal play of life.

Sitting in the corner of a crowd, I was looking at the animation of life. I wondered, was I looking at the animation of my life or their life? It seemed like we all were searching for our recognition in nature or you may say we all were realizing our identity with each other.

A lot was being neglected and a lot was being taken care of seriously, including a lot of what was precious, a lot of what was useless, and a lot of what was stuck in struggle.

Who has achieved his recognition and who was fully recognized by others and whose recognition was still stuck in struggle was not properly fixed so far. This cannot be estimated even from the earth drenched with blood.

It is not possible to stop a storm before it begins blowing. All have to bear the aggression of that. It is not easy to fix the boundaries of a storm. This can only be estimated. No restrictions can be fixed on the will of a storm. Otherwise, it looks strangely calm before a storm starts blowing, but the heart becomes little bit uneasy, birds become restless and start crying, dogs start barking and other animals become frightened. Birds and animals note the symbol of a storm before the blowing starts. Only man sheds tears after the storm starts blowing.

This all happens due to the selfishness of man. He always ruins his present under the burden of the past and future. But animals and birds don't do this. They always keep themselves in touch with nature and depend upon it and keep in mind only their present.

Sitting in the corner of the crowd, I am looking into the philosophy of life. Every life was acting its role in the play of life in which the main role was that of man. You may say he was the hero of life, maybe in a negative sense, too. Man's power was prevailing everywhere. Man was able to do anything. No other species seemed powerful enough to give any direction to man but otherwise all were bound to dance over the tips of man according to his wish. Man concentrates on birds when he thinks about his lust and concentrates on animals when he thinks about wealth. It shows his animal concentration that he never tires of shedding tears after experiencing a loss. The thinking of birds doesn't accept boundaries and never allows its flight to be constrained by boundaries. The cage is also a symbol of man's concentration of animals.

To show his sociality, man concentrates on birds for his lust as animals generally abide by the rule of nature for sex, which is not acceptable to man at all. Animals never perform unnatural sex but wait until the female comes in heat. They abide by the rule of nature and don't practice sex only for enjoyment. But man, by calling himself a social animal, goes two steps ahead and ignores the rule of nature by which animals abide. Otherwise, when he saw that birds live in the pattern of a pair and abide within natural rule, he immediately jumped to the other side and started to love his pets, which are animals. Could it be that man thinks that it will be good for his socialism to stay away from nature?

I was sitting in the corner of the crowd but this means that I was not trampled under the crowd. Actually, nobody was taking notice of me but I was not separate in any way from them, who feel proud by connecting their past with animals—specifically, with monkeys. It seems as if the first wise step man took to come out from the crude contrivance of being social took birth from his will to be near nature. It seems to be the result of some selfishness of man that he has become free from the explanation of the theory of the direct origin of man. Now, he is totally free from the need to determine how he came on the earth for the first time. Woman came first or man or both came together—he knows now it doesn't matter. These talks have now become useless for him. He very cleverly put this onus on monkeys as a way he could explain how he come on the earth for the first time. It seems man has become social by development from the monkey. The developed form of a monkey has not yet reached up to the sun and stars but they have discovered that the earth is the broken piece of the sun and the moon is the broken part of the earth but who asked them to rotate and revolve around each other is yet to be discovered. A monkey is still a monkey because he is not interested to discover anything about how things are made, and when and why he came into existence on the earth. He is also not taking any pains to think about the history of his elders and it looks like he is not interested in joining the social brotherhood any more.

This worry, which is only that of the developed form of monkey, is a means of man in order to establish the sway of his thoughts to make more discoveries. The developed form of monkeys will not allow other monkeys to be developed anymore, because it would be an insult to his status to allow anyone else to join his developed community. Already, more than enough monkeys have generated their own developed community, so they are feeling the dire necessity to kill other monkeys with bombs. Some of them are killing with hunger. So now, it is not possible at all to allow other monkeys to join the developed community. They may raise other types of new difficulties, so it is not possible at this time to take any risk. So, under no circumstances, can any opportunity be granted to the remaining monkeys to become a social being. At this stage, it is also not possible to arrange an education system for them. There are also many other problems associated with this. However, the question whether some men might again be made into monkeys is open and serene efforts for that purpose may be started.

When man was in the process of taking on a developed form from a monkey, at first he started describing himself as a social animal so as to break the relationship with other animals. He had no confidence over his own mind and otherwise to show the best shape of his sociality and to appear different from other animals, he invented cloth. Now he was looking different and more sociable than his ancestors. Now, whoever was wandering naked would remain a monkey and those who were wearing clothes would be the social being. Clothes became more or less formal in various places, but it became necessary for people to look sociable.

I have heard that our ancestors did not know how to burn fire. If they have come to know how to burn fire, then they must have taught the lesson to their changed form. They must take revenge, which is why they do not allow the remaining monkeys to be a man.

Otherwise, what has man achieved by learning to burn fire? Then what if he cooks his food with fire with one hand, but with the other, he gives bombs to others to eat.

First of all, when man became socially distinct from animals, he forgot his origin. He controlled the life of animals and put them to live on his mercy. Now life was fully dependent on the mercy of man. Birds could save their life a little bit due to their wings. When man tries to capture them, they fly in the air and save themselves before coming into the clutches of man.

There is animal, bird and I am sitting in the corner. When I look at myself, I am looking very well social in wearing clothes. Animals and birds are also wandering hither and thither without any clothes. I saw that my clothes may be very beautiful but these are a curtain between me and nature. I saw birds and animals are very close to nature. Their behavior towards nature is more meaningful than mine. Under the burden of sociality, my behavior towards nature is cruel. Realizing the nature of manhood, I was compelled to shake off the dust of clothes and, realizing the nature of sociality, I was compelled to ignore the surroundings where nature was prevailing due to ill will to put the stamp of my behavior on it.

Freedom

At the time of my birth, at first air entered in me and I started crying. My cry was the sign of my good health. If air does not enter into any part of the body, it means that part was not fully developed due to some type of deficiency. The effect of this is that the part becomes defective. If air doesn't enter, it means the person is disabled from that part forever. Non entrance of air means that air has not given any pain, and without pain, there is no crying. If somebody doesn't feel pain, that means he is not healthy. So my crying was a very good signal of my good health.

So my relation with pain began from my birth. Not only for me, but the truth is that the relation of every man with pain begins with birth. Even which language my tongue learned for the first time was that of crying. Maybe this was presented to me by nature, which was related to my health. After that, which language doesn't remain as natural as the original first one? After that, comes the language coated by the path of life, home, society, country and good and bad deeds in which apparently appear the effect of the lessons of doctrines. Whenever the matter of the sufferings of life comes forward, the name of man comes to the top of those forces that create the sufferings for man and the name of nature comes too far behind. Which pain was given to me at the time of my birth by air that was necessary for my health?

Then I started forgetting the natural language and started to learn the language that was prepared for me, keeping in view all of my necessities. That language was taught to me by thinking about what was necessary for my knowledge. That knowledge became an important part of my life. Slowly the coating of that knowledge became much deeper by the passing of time. My surroundings and ambience were becoming my recognition and identity. My family, tribe, collateral, heritage, brotherhood, village, town, city, country, and world all started becoming part of the share in my recognition and identity. This was a very important time for me. The main change during this period that came was that my relation was being broken with nature and I was being surrounded by worldly things and was becoming totally materialistic according to my needs. The coating of polished rites and traditions of cultural, social, and familial connections was covering all my naturalness.

Confronting the realities of life, I started forgetting the blessing of nature. It became my habit to find the meaning of life from that atmosphere in which I was being raised. Every day, I was being buried more and more under the materialistic life. It was my first defeat before nature, which I accepted in my innocence. I was giving up naturalism and becoming a puppet in the hands of selfishness. Slowly I was feeling good to be the puppet in the hands of my elders and I started growing up in the atmosphere established for me. By accepting all their rites and traditions, I was growing to be like their hope of the future. Maybe there was no other way for me. The chains around them slowly surrounded me too and I could not feel anything at the time of accepting those. These were the same iron chains that were made to control the elephant, keeping in view his strength. Their ecstasy can be crushed and can the strength of the elephant become the meaning of slavery? I was never able to remove this defeat from myself. It was possible to ignore the effect of this defeat but it never happened and all the time it ran with me as a sign of slavery. I bear the insult of this slavery all the time. Maybe my elders also learned this in the same way. They were also coated in the same way and they were not able to understand. I was a natural creation only in name but actually, I was living as a puppet in the hands of rites, traditions, laws, and rules. It was looking now like there was nothing of myself in me but I have become the will of others. Now my struggle was not to remain natural but to maintain that status whose back was bent under the burden of life and cannot be straight ever again.

Then I started to learn that language, and with whom I had to learn that language too, which was necessary to teach me to bear atrocities and I have started to bear the burden of that too. That was also necessary for the struggle, which I have to make to be alive. That was also necessary to remove the hunger of my stomach. Maybe my elders also removed their stomach hunger in the same way and they must have thought this path is right for me too. To fulfill my daily needs, to make progress and enhance the inheritance, to keep account of it, to understand the science in the age of science and to take advantage of it, to measure the earth and sky and to understand it, and to get every other knowledge which comes into my share from the knowledge of the world, I made every effort that I could to achieve it and I passed through all those paths of pain and pleasure, which eclipsed me altogether and so my relation with nature was totally broken.

Not only was I separated from nature, but the people who were in possession of the resources of nature fully introduced to me their identity and power. Now I was that horse on which they put a saddle so they can take a ride at any time by their own wish. My relationship with nature has been lost far away in the footpaths, which I was unable to find even if I wanted to do so. Even if I become able to do so, I must find my life snagged in thorns and some private owner must have obtained the order to finish at his own wish. Every resource of nature was stamped under someone's ownership and life was a slave of those stamps. My struggle became limited to the removal of some stamps and to jot down my name on that. This was the same struggle of my elders and maybe they thought out of their experience that this is the right path to engage me with life in the same way. I accepted all that silently, with the slight annoyance of a calf who, as it grows, after a little obstinacy starts to follow its elders' instructions.

After I learned the language of crying, my elders must have selected this path as the right one to remain in, which they must have obtained from their life experience as they struggled to have a cultured life. Without this language, it was not easy for me to remain in the struggle of life. However, my mother has told me about the freedom of birds many times. Actually, she not only told me about the freedom of birds but by connecting and comparing those with the limitations of human life, she also taught me much more. I don't think she has less love for nature than anyone else but her affection for her son must have compelled her to make my back strong. Keeping in view her life experience, she made me strong with the bitter reality that nature alone is not enough to live life but the materialistic life is more important to understand. She was well aware of the aspect of manhood, socialism, and materialism. If she was aware of the freedom of birds then she must have had the knowledge of the destitution of animals too? She must have been aware of the importance of social life so she must have taught me those paths, which she must have felt were right for my struggle of life out of her experience.

I have also heard the songs of birds so many times. Sometimes I think that the proper language of birds and animals must exist too. Cows, buffalos, and their calves easily perceive each other from far away by their voice even if they are not visible and become able to recognize each other. Cows and buffalos become able to recognize the hunger of their calves from their voice. It is common that animals understand each other's voice but their language is limited to their needs and undeveloped when it comes to their limited freedom. The songs of birds even impress human beings. Sometimes the meaning, rhythm, melody, and cadence in the songs of birds so convinces man that he starts to copy them. Sometimes man copies the sound of movements of the feet of camels and horses during their walk and copies the rhythm they make. Sometime man was so impressed by the walk and the sound of a string of bells around the neck oxen full of rhythm. This is directly linked to the movement of the heart and is a tribute of nature.

Birds are generally free from the grip of man but animals are generally held in the grip of man. Maybe this is the difference which doesn't remove the naturalism from the songs of birds. Birds aren't coated with language like man with strict laws and rules by which they have to accept slavery. They are not buried under the tradition of family, tribe, or socialism. Their naturalism is their strength and allows them to enjoy life more than humans. Man is so materialistic and he indulges himself in these things so much that he stays away from the treasures of nature in which there is enjoyment. His materialistic wishes don't allow him to come out. I am very impressed with the easy life of birds and disappointed with the difficult life of animals that are compelled to live in the hands of man, the social animal. They are not able to keep their destination away from the earth where the social animal is completely in possession of everything. Animals and the poor are the main victims of this possession. The pitiable condition of the poor and animals clarifies their whole story so sometimes the freedom of birds becomes my ardent desire.

The coating on me realizes another important thing that put the stamp on the wisdom of my elders on me such that I was still one step ahead of animals and I was called a social animal. If this coating separated me from nature, then on the other hand it brought me out from the line of animals and gave me the knowledge that I was able to put forward the struggle of life by using the knowledge and became able to break the chains that were put around me to finish the naturalism. Even if I failed to do so, I can put forward my unfulfilled desire to the next generations. This desire is not for me only, but this may be for my elders too, so they might have suggested this path for me.

I was feeling that maybe I am ahead one step ahead of the animals but I am one step behind the birds. This was because the earth is also the last destination of me like animals and I was also unable to fly in the sky. This is the main reason that the race of animals and the poor ends soon and comes into the grip of materialistic man. Maybe the change of season has convinced millions, billions of bevies of birds like sparrows to leave one place and to go to another place. So for unaccountable reasons, these sparrows were going from one place to another. I never before saw birds in such large numbers migrating like this. Many questions came into my mind. How can they come together in such a big number? What will be their connecting source and how would that have been working? How could their consent have taken place? How could they have decided to go like this? How could they have found the resource at a big enough level? Why were they migrating together in big numbers like this? What is their system of cooperation and how could it have been working? There were many other questions in my mind for which I had no answers. Do they have any perfect language or not? If they have any language, then of what type is it? I have no answer for this. Otherwise, I was feeling the fragrance of freedom from their flight, insouciance, unity, and songs. I also started examining the meaning of my freedom from this fragrance.

Witness

They started looking for justice. A suggestion came forward to agree upon god as justice. But this was strictly opposed.

"How can this be agreed upon? We don't believe in god. Those who believe in god, they think god is so powerful that nothing happens without the wish of god. But if there is any god then he must take action on atrocities of man by man. Some people are living in hell and on the other side, some people are living in heaven and suck the blood of those whom they have compelled to live in hell; is not god watching this? If there is any god, he must come forward to finish this injustice. Man is not only the victim of man but he is the victim of natural disasters too. Even otherwise, if there is any god, can it be denied that he writes in the fate of the poor to be born in the home of the poor and for the rich to born in the home of the rich? If such a thing doesn't exist, how we can agree to admit the justice of such a god? Try to find another justice," some of them said.

Many thoughts and suggestions have been put forward but none of those was acceptable to all. A suggestion of election was also put forward but keeping in view the danger to the unity, integrity, and impartiality and to avoid the creation of any unnecessary opposition, that idea was also not accepted and rejected vehemently.

"What is the need of justice to truth?" one of the logic came forward.

"Which truth is not hanged? Even truth is hanged by justice itself. Another matter is the fact that truth doesn't die. Even if for the time being it looks like truth has died, it takes no time to be alive again. As some people say, life is false and death is truth." The logician was taking another turning point.

"If you think the truth is being crucified by justice himself, then why are we searching for him? If time has to decide the matter, then we should leave the decision to time. We should not accept any other as justice," the Logician raised its head.

Many were dozing. Many were not paying attention as they were of the view that the majority decision would be right and acceptable. Those who were actively taking part and were very vigilant also thought that it would be better not to oppose this view rather than to go without any decision.

Somebody took the benefit of silence to conclude the matter and to take the consent of the others and shouted the slogan, "Time is the justice." All were willing to conclude the matter; some totally agreed, some half and some did not agree; some, out of respect to the majority decision, became committed to the logic and no one raised any further objection.

A superabundance of logics, which were to be produced before justice for decision, now started to appear. There were a heap of logics. Logics were being disregarded on a large level. Logics were being neglected everywhere. The position of logics became so bad that now nobody was ready to talk on any logic. Now, nobody was ready to speak in favor of or against any logic. Which logic is likely to be admitted or which logic is likely to be ignored or which is likely to be amendable, nobody was ready to say, as there was a possibility of danger to the unity and opposition may raise its head. There was no beam of being united that is able to spread itself like the shining rays of the sun and moon. All were looking towards their goodness and they were not ready to lose it without any solid reason.

It was the reason that there was no opposition coming forward. A strange stagnation occurred.

Time was feeling disgusted but was watching the activities silently. One of the logics raised its head like a mushroom that emerges out from the soil after a wet atmosphere, and said, "Consider the whole earth as one unit. Finish all the malodorous actions of the past. Finish all the rights of ownership, finish all the lines of boundaries of the countries, prefer the importance of the human being, let his rights be accepted under the laws of nature and give another opportunity to live together with each other."

A loud noise of thunder was heard but the sky was clear and there were no clouds. The thunder was not from the crash of clouds. Actually, the power came from another planet, which had finished the logic from the beginning. It looks now as if man can never sit together in the future, as it was not acceptable to the powers of other planets as it may create danger to them.

Now new questions took form, that is, the power of the earth and man versus the power of other planets. There were new questions gleaming over all the faces. Nobody was looking towards the dead body of logic. Actually, everybody was

308

so engrossed in the new threat that all were ready to put the dead body of logic in the deep grave, as it would not take birth again. It looked like nobody had even a little bit of sympathy with the dead body of logic. Some were responding through the heart and some were apparently looking very happy. Some innocents were looking hither and thither feeling strange.

One tired voice was trying to say something out of the heap of logics. It seemed to have a clear deficiency of words to explain its opinion. Maybe there is another reason that it was not able to explain but the voice could easily be heard.

"Logic is not a shine or shade which is to be felt, it should have a tongue to explain clearly," someone poked.

"Human division should be done effectively and deeply. If that requires the immolation of some men, do not hesitate about that. History is the witness that divided society so it can achieve more progress. Try to understand the importance of the quality of poison. This is necessary to finish the divagation of man. Be legalized to grow poison and to embitter. Bombs should be treated as the only way to peace. All speech against the rain of bombs should be declared illegal. The breast of the voice of human rights should be searched with guns and bayonets. Progress should be loaded on the heads of people to be kept safe in a few hands under the supervision of atom bombs. There should be more infernal pits in the air and sky. This is the only identity of time," Logic apparently unveiled.

All appreciated the courage of logic. All were remembering their real faces. All forget for what purpose they have come here. Without hesitation, there were long lines to sign the logic. It was difficult to understand: why were they doing backside towards that after signing? All were continuously signing and doing backside towards that. At the first instance, it was felt that they were saving the secret of the logic but after sometime, it could be seen that they are not hiding anything and otherwise they are openly in favor of saving the logic. It was becoming their power show. Now there were so many backsides towards logic that nobody was able to reach up to the logic. Now they were sending their messages of acceptance. People standing doing backside were doing this work hand to hand. Now logic was not visible; only people can be seen standing around it.

Nobody could imagine that there is any logic behind the people standing there. You could only see these people. Their decisions are now logical. They made their faces so beautiful; could it be that the logic behind them was also looking attractive, as someone covers their inner lust with beautiful words? As an inner blackness someone hides in a beautiful face? As rich always remains wise. This logic became the reality, and it was traded and logic grew up on the rivers of blood. Logic was lying behind it like an octopus. People were becoming attached with its tentacles. Some were being squeezed breathless in its twisting grip. Some were declaring it to be heaven and some were saying it is a hell. Some were habitually able to bear this. Some were compelled to bear this and they have no other way to come out of it. They gave the name to patience, of tolerance.

Some people were worried.

"Worry is equal to a funeral pyre," they were being assured and, being injected of this assurance, were able to sleep deeply.

Time was laughing boisterously. "I was never alone. The soil of your past, today's courage, and tomorrow's hope are the witness before me."

Your Problem

It may be my story. Although I am not different from you, my story may be different from yours. I understand very well that you are only my entity and I am not able to keep my entity without you. I admit this thing, that this is not my story; actually, this is your story. Everything of mine is attached to you, so even if it is my story, it will become your story. You are my existence. That is another thing, that at some point you realize yourself trivial. So I want to keep you vigilant right now so that when you realize this, you should not be disheartened. Actually, you are the main part of my body. We cannot see the importance of any part as either less or more. The whole body is important. We cannot go away from truth. The truth is that, as independent as I am, without you, I have no existence. If you are my existence, then it is your story. I think this is not the issue of an argument. This is for you to decide. You can decide whatever you want. Actually, I have no need to decide this.

Have you ever been face to face with yourself? Keep in mind that when you are going face to face with me, you will be face to face with yourself, too. You could say that you are going to be face to face with your thinking.

Do not think that I am going to give you any decision or suggestion or am going to serve anything. I have no such new knowledge that even you do not know. Do not be so foolish as to treat it as a final decision. I am not indifferent about anything. I am your identity, way of living, and the sign of how you are passing through the world. There is nothing of myself in me. This is what I am saying again and again, that you do not care about me. You are going to be face to face with yourself. When you go through the surroundings, you will become my shape. I am only saying that which you are feeling.

I was born with the development of your sagacity. You were also living like ordinary animals. I was not needed by you very much at that time. You were also behaving like the animals. You were free and not being controlled by any law or any other restrictions. That which you thought was right, that was the law. You only abided by the laws of nature. There was a very big difference between other animals and you. This difference was even from the birds, too. Your brain was sharp and intelligent, so it started developing fast. Your developing mind gave me birth but greed also took birth at the same time, which introduced your relation with selfishness instead of nature. I was dying compressed in it. Greed was never my favorite. But do not think that I am trying to break my relation with you. Actually, this new trouble, which took birth in you, made my path difficult. Here the question of status arose between me and you. This trouble had no effect on me at all but it started making a difference in you through classification and gradation.

I am not trying to tell you any ideology but this is a truth, which also started to be written in the shape of history by your developed mind. How many of your generations have been finished up to now? This is also your deed and, about it, I have no idea. Man also comes on earth like any other creation and dies like them, too. There is no eminent or indigent being which is not mortal. There is no effect on me of these incidents. I silently enter in your next generations and remain in the present in them to do your last ceremonies and to move them forward. I will be finished only with your total disaster but if you are there, I shall exist.

The developing mind opened the new paths of progress for man. How many species have become extinct during this progress? This is no concern of mine because I am concerned only with you. This new problem that took birth in you wrote a new story of destruction of man by man, which became more pitiable due to his developing mind. In every age, man exploited man and created carnage for humanity. Every evil developed in all ages as man did his best to capture the resources of nature in the name of progress. It is not my concern how man has passed through the ages. He has written this history in books in which you can read the stories of my destruction, too.

During his development, the mind of man stood before the evils of man. When that evil was being laid in the cover of goodness, they uncovered the evil before the people. I saw my decoration in the ideologies of theism and atheism and also saw suffering in the same ideologies and in being neglected. Man sacrificed for man again and again. Blasphemy raised its flag again and again. Above the natural needs, man became the trader of nature. Man became busy fighting wars and traders became too busy to trade.

The developing mind's progress started to capture natural resources. With the propensity of capturing and wishing to enter the victory over nature, man took that path on which he started to ruin nature. He started boasting of his progress. The race to capture the natural resources made him so cruel that even if he felt any danger from any species of earth, at first he tried to control them and if he failed, he removed that race from the earth forever. He tried first to control man, and if he fails, his destruction is also determined. The system over the earth is that the destruction of man is determined by the hands of man.

To make sure of the sway of trade and to snub the public memory forever, they invented nuclear bombs along with other, routine bombs. Could man now be massacred like carrots and radishes? To save the booty from the others' wrathful eyes, they made countries with countless power. Countries became poor and rich. Everywhere, man was becoming the victim of man. Trade was brought under the laws and made legal to crush the common man. Power made paths easy for purposes of dividing the human being. In the name of humanity, the massacre of humanity was made permanent. Only crocodile tears were shed over the tragedy of humanity.

Nature was crying but to make its mockery is the habit of powers. Man's ego became so strong that after getting the blessing of nature, he treated nature as trivial. Instead of taking in view his own wrongs, he is considering to accuse nature of being like criminals. Under these compelled circumstances, frustrated nature is watching to see whether man can finish his sagacity with his own polluted noise of progress. But man has no solutions for this self-created dilemma. Man is not creating problems and harming nature only, which may destroy earth and water, air and fire, while every natural thing over earth may be out of control and finish everything, but he is also creating a dangerous situation for the coming generations.

Sometimes I feel danger from my own definition of why I am representing only one species over the earth and why I am not a part of all the creation when man is creating the danger for all of creation. I was thinking that man could live only in love as male and female like other creations over the earth. But his savagery has damaged my whole image over the earth in the eyes of all the species of the earth. Now I am alone and, of all the species upon the earth, appear to be the only ugly one.

After tying my existence tightly with religions and putting in it their own ill will, man tried to fight wars. Of thoughts, which at some time were made for humanity, man is trying to make the platforms of wars. This is my compulsion, that all good and bad things are my parts.

My existence cannot be kept limited within the boundaries of countries, which exemplify the selfishness of man. My existence is nothing without you but I am not limited within boundaries like you. My existence cannot be kept tied with religion. Actually, selfishness cannot limit my existence in it.

It is actually wrong that I try to tell about my greatness. You are my existence. Society is the name of your existence and unity. The strong health of society depends upon your strong thinking. No power can take me in its grip. Power is not the symbol of my existence. It is collected by man's ego. Power is not my foundation. Otherwise, my foundations are those people who bear the atrocity of power. But the question arises: Who becomes the victim of discrimination? Who becomes unable to get the blessings of nature, which are in the hands of power? Nature never discriminates against anybody. All are equal before nature. It is man who prevents others from getting the blessings of nature. The laws of nature are the same for all. My preceptor cannot be a symbol of power. My preceptor can be an ordinary man who put his head on another's hand or kisses the crucifix of others.

How long I have been looking for the ages of changes and up to when these will continue, may not be in your control to fathom. Which shape you want to see of me, it is also your problem. What I am now, it is all your contribution. I will always follow you in the future.

End of 2ND Book

3RD Book

Nature Summons

(Jas Fiza Science)

By:

Jas Fiza

Nature Summons

(Jas Fiza Science)

By: Jas Fiza

Publishing Tear: 2013

322

Nature Summons

(Jas Fiza Science)

Translated from

Punjabi Version

Javab Talab

By: Jas Fiza

Punjabi Version Written By: Jas Fiza

Dedicated to:

The Whole World

Preface

Deep study of Indian philosophies and several sacred books like Vedas, Shastras, Upanishadas, Purana, Ramayana, Mahabharta and many other books gave me deep knowledge about different prior ages. But I grew up from the mid of 20th century with the development of science. I remember when my father brought a radio that ran with a big battery for the first time into the village. I remember how it caught voices from other places far away from our village. It was very surprising for the villagers who came daily to see and hear this miracle. The same thing happened in the life of my father when our grandfather brought a car to the village for the first time. There are many differences between cars of 1932 and cars of today. These revolutionary developments of science are very notable. This growth also developed very quickly. I saw radio to TV, computers, satellite, missiles, airplanes, drones and every development that is at present you are enjoying. Comparing the developments that I read about in the books of Indian philosophy with present development, I have come to some different conclusions than the concepts that currently exist.

Reading those books and in looking at today's development and advancement of science, I saw that it seems as if we are repeating history again. Where are we going? What is reality? What we are doing? What was that and what is this? Are we going right or wrong? What we can do? Do we have options? Do we have an option or no options? I view answering these questions as a new science.

I have named it 'Jas Fiza Science' because this type of science has not existed in the world before. This science has a new ideology to believe, and there is much work to do before real conclusions are made. I think it is the starting point. This is my thinking, and I am not stuck to thinking that this is the final science. However, I will stick to it until I have a strong reason to change my mind. I always look forward to welcoming new ideologies that force changes in the existing concepts with a reality that is widespread in the universe. Then, we expect that we have knowledge of that truth.

This is not my first book on this concept. I have written a few more books on the subject. 'Execution of Time' is a novel, and '2nd Moon' is a collection of short stories on the same subject where I adopted the same concept. 'Sabadan da Chola' short stories, as well as 'Javavdei' and 'JavabTalab' poems, are written in Punjabi script on this same topic. The present book is a translation of 'Javab Talab'. I think that I have done the complete work from this side, and I have nothing more to say on these topics.

Jas Fiza

Nature Summons (Jas Fiza Science)

Accept my presence

To face the truth,

Or you may think

I have come.

Otherwise I never tired

Answering your questions

In your wisdom

I have seen

From ages to ages

Flirtation of your wisdom

Don't be inadequate.

In this age too

To throw the bomb of wisdom

Use your cleverness

Up to any level

In the present age too

Don't live in doubt.

Use all your efforts

Or efforts through science

To see the result

Of your doings

I have seen your behavior

Even before this age

You never did less

At any age

You made all efforts in the past

To be the creator of nature,

But you were disheartened all the time

Broke the back bone

Of your knowledge,

Put your existence in danger.

I remained present there

To save your reverence

It's not in your control

Or your knowledge

To compete with me

In this age too

Use your all efforts

Or your knowledge

Of your power

Of your possession

Of your existence,

Use your knowledge

Or things made by knowledge.

Fulfill your desire

This time too

Looking very eager

To become creator of me,

I can't stop your eagerness.

Vanity made a home in your desire

I am totally unable to stop

The vanity of knowledge

Which you are giving air to yearning,

Making castles of knowledge

To take me in possession,

To take charge of my rotation

To be my driver

Or my creator

But I know fully

And understand too,

 I will tell you

When the time will come

And will show you

The glimpses of your power

I nourish living beings,

Vegetation and plants

On every part of the earth

I am the creator

Of every living being,

Vegetation and flora

I am the driver of earth

And of every flow of earth;

Of your every path

From birth to death

You may recognize it

Or you many do carnage to human beings

Or suck

The blood of human beings

We have open competitions

To test all of this

However I will, time to time,

Answer your cleverness,

Reckoning the account of your doing,

Answer your vanity,

Tear your mask

You are proving cruelty

Noble cause for the world

But running your business,

I will break your utensil openly

Mortify you openly

I will certainly clarify

How you are doing.

Noble cause for human being

In every age we faced

In front of each other

Making exceptions

All the time with each other.

You stumbled all the time

With your own knowledge

And the vanity of your knowledge

From the cause of nature

By making laws of earth

Full of your selfness

Full of your ego

Show of your vanity

Exposure of your power

Vanity of equality with nature

Under the shadow of guns

But glints of your greed

Uncovered your face

Full of your selfness

You could not recognize yourself.

Always forget yourself

Your existence as a body

 Five elements of me

And created by me

How can you win yourself?

How can you win your body?

How can you declare victory?

Over your breath that is air

Over the blood of your body;

Or over the body that ultimately

Will be the part of the earth

All your five elements

Is the trust of me!

How will stand equal to me

How can you get victory?

Manikin made by me

He will make foolish

To his own human

What will you teach me?

You have not the courage

To achieve victory over me

To have equality with me

In every age you lost,

You shattered in every age

With your knowledge

The vanity of your knowledge

Must be defeated

In the present age, too

Wait for the due time.

As a matter of fact

You are in pitiable condition

When you make a claim

To be my sympathizer

High pitched wailing

To save the earth,

Put forward the warnings

Proclaiming before the world

Showing your intelligence

Show the fear to the world

Of global warming

Shortage of water

Melting of the glaciers

Calamity of global warming

Over the earth

On every living being

Forgetting your own doctrine

When you said openly

That earth is the part of sun,

Broken from the sun

Became cold slowly

Livable for living beings,

Vegetables, plants and flora

When the earth is becoming cold

From millions and billions of years

How quickly it changed processes,

Which method compelled

Which thing enforced this?

 Some other sun

Or the present sun,

Pressed the reverse gear

Started working in another way.

Or the earth changed its process,

Started working in reverse suddenly

Or man is accusing himself

Of any of his own blunders

For which he thinks

Earth is being compelled

To change its process

The earth that was becoming cold

From millions and billions of years,

Now in this century

Started working in reverse.

Now global warming is coming.

Does man not think?

He is making a mockery

Of his own theory,

Of his own doctrines

Does man not feel strange?

By talking to the contrary

He doesn't understand

The rules of nature

Or his work is only

To throw a spanner to work

Or to shut his eyes from nature

To show the power of science

Or desire to show his wisdom.

Man is on the wrong path.

He forgets all the time

His own existence

He forgets always

The creator Mother Earth

He forgets every time

Mother nourishes him,

Nature is his existence.

How can he possess the whole?

Of which his existence is part?

He always forgets

I am the heat of his body,

It's me who moves him

From his birth to death

My effigy expecting

To be my monitor

Forgets all the time

The power of nature

Forgets what is in me.

Millions and billions of earths

And suns are my entity,

The beauty of my sky

This sky is mine.

Earth nourishes creation

Sun helps in nourishing.

How is it possible

Earth will be warm again

Like the sun reversely,

Earth is becoming weak

Earth is becoming cold.

It may happen

It may have

Some defect in equilibrium

In the stability of earth

Or have to take some steps

To cure its process

But it doesn't need

The help of man

Or the sympathy of man

Trite wisdom

Lack of sagaciousness

Boast of sham knowledge

Fake claims of wisdom

Hopeless ego,

To be the creator of nature

To be the great

Show of power of knowledge

Bad smell of unfulfilled greed

Tale of partiality

Story of atrocity

Blemished deceitful intention

Unable to face truth

Having false notions

Earth is fully able

Nature is so great

Able to cure itself

Able to correct

 Human wrongs

Who weakened the earth?

From inner to outer,

Using the energy of earth

For his luxurious living

Without any necessity,

And giving the challenge

To the great nature

Earth may take a deep breath

And give birth

To another moon

It will be shattered.

Man's utensils of knowledge

Will come once again

Of the same undeveloped stage,

Once again will start thinking

For new ways of progress

It is not easy for earth

To cross through this process,

To get back the strength

In which it maintains

Its own equilibrium

While rotating around its axis

Revolving around the sun

That never comes late,

Even a second or minute

Working on its rotation

For millions and billions of years

On regulated speed,

Never felt a need for the help

Of any human being,

Or human knowledge

Or human science

Crocodile's tears are false,

Man's knowledge inconsistent.

Vanity will never win.

Man cannot possess

Any element of nature,

Cannot lead to any element

Man defeated in every age.

He won't admit his defeat

Never forgets to show

Ostentation to nature

To possess Mother Earth,

To search the stomach of mother

Declares himself the owner of mother,

Again becomes the soil of earth,

Including his knowledge

Sometimes man makes mockery

Of his power

Of his vanity

Of his science

Of his knowledge,

Brags over progress

Forgets its familiarity

Becomes showy

Of trivial power

With toys of science

Proclaims in the world

Of wonders of science

He looks to science

As a symbol of power

Symbol of his strength,

Brags before his mother

To show the whole world

Claims himself to be

The guard of nature

Mother smiles over son

Remembers the days

When man was in diapers

Weeping and crying,

Sees the foolishness of son

Looking upwards towards the sky

Watching the sky full of stars,

Does the guess work!

More suns, more earths

Revolving in the sky

Speculate about God's living

Not living in his heart.

Must be living in the sky

On some other planet,

Forgets everything

Upwards downwards is illusion

Sky is everywhere

Up, down and all sides

Man speculates over other planets

Other suns and Earths

Like his own Earth

Speculators have made progress

Must have developed science

May have advanced more than earth

Then entrapped himself

In the false notions,

Speculate on their power

Of some other planet

Must evade Earth

By more powerful aliens

Proclaim to be aware.

Alert his scientific army

To fight back aliens

Now Earth has become

A meek spectator,

Becomes very weak

Foolish son has grown up,

Will guard his mother

From the dangerous aliens

Of some other planet,

Praise narrating his vanity,

Give shape to his dreams

To get admiration from mother

To save her dignity

See how the mock dreams

Of his greatness

Challenge the world,

Show his intelligence

And acts of his kindness

On unfortunate nature

Takes big decisions

To fight on behalf of nature

For the safety of nature

To defend the star war

And see beautiful dreams

To win the star war,

Give a pat on his bosom

For the achievements

Made through his science,

Brag about his greatness

Over nature

Mother has nothing to do

Except to smile on foolishness

By hearing the brags of her son

Mother can't explain its destiny,

Bears atrocities on its bosom

Bloodshed by sons on sons

Distressed from her sons

The whole creation perplexed,

Carnage everywhere from fighting

What will show more miracles?

To raise mouth upwards

By absorbed speculations

What purpose will it solve?

To whom he is telling

These foolish speculations

His devilish talking

Talking of his own cleverness

Looks like he has lost

Power of balance, thinking

In the vanity of power

Malodor of bombs on Earth,

Wants to spread in the sky

Wants to pollute the sky,

Taking evil dreams

Making the Earth fire ball

Making the Earth dangerous,

Now showing himself capable

To be the guard of nature,

Forgets his stupidity

Before nature

Showing his power

To his own grave

Ignoring the problem of Earth

Ignoring the problem of Nature

Talking about the star war

Showing the danger of

Aliens of other planets

Showing himself smart

Wretched man!

The vanity of humans

On the progress of science,

Believe the satellite system

With the cameras and telescopes

Making very distinguished

His appearance and identification

Writing a different history

With Bombs and drones

Feeling very happy

About the material of destruction

With the accumulated power

Of knowledge plus knowledge

Giving lead to Satanism

Raise mouth upwards

To proclaim his knowledge

Of dangerous materials,

To tell about his miracles

Explaining his knowledge

As very comprehensible

Of a sky full of stars,

On his distinguished discovery

Of dangerous black holes

Of which Earth has to cross

While revolving around sun

Shows fear of black holes.

If Earth is caught in those,

Everything will be finished.

Then started vaunting

Over his unfathomable knowledge,

His countless greatness,

Thorough searches of sky,

To understand all the ways

Of the work of nature

Now leaving no chance

To tell the whole world

To delude the human being

In the conceit of power

Man who is expert

At shutting the mouths of people

With the language of bombs

To flow the river of blood

Drenching the earth

To downgrade the earth

To show the earth pitiable,

Show his ill-bred

Whom he is making a fool of,

Whom he is frightening

Has he now started thinking?

In the future he will show

The path to the earth,

For the safety of earth

From the dangerous black holes

Or he will give new directions

To revolve in the sky

In the future he will guard earth,

To be declared Messiah of earth,

Take the charge from nature

To rotate and revolve the Earth

To run the work of nature

What is in his mind?

Earth has become indolent

Unable to rotate or revolve

Is this what he thinks?

Earth has to depend upon man

Or has to take shelter of science

Where was his science

From millions and billions of years,

When the earth was performing

All of its duties accurately

Showing up to date

All colors of nature

Rotating around its axis

Making days and nights

Revolving around a sun,

Made the seasons,

Performing its duties

Never reached late

Not even a minute or second.

What meter keeps accuracy?

What engine runs it?

What is so accurate!

Who is the administrator?

Of such a big system,

Running without any interruption

For millions and billions of years

Why is man anxious in this century?

About rotation of the Earth,

Of which system

He has no knowledge?

From where will he bring?

The substitute of that

Will he find it from the sky?

Or by searching nature

He is unable to search

The abdomen of his mother

Whom will he make foolish?

What does he understand?

Why the Earth is rotating

Around its axis,

Making days and nights?

What is the relationship?

Between light and dark

How does it give warmth accurately?

To the creation over earth

To every lively hood

According to his need,

Borrowing from the sun

In its regular rotation

Without the help of a human being

Or his science

If earth stops for a moment

Or nature stops for a moment

All will be devastated.

Whether science has power

To make days and nights,

Whether science can bring

Different seasons and weathers

Who is able to stop it?

Whirlwind, twister or tornado

Unable to stop an earthquake

Or to stop tsunamis

Is science able?

To make Heaven from Earth

With his few toys

Engaging human beings

By capturing natural sources,

Playing cat and mouse games

With human beings,

With the illusion of money

Made from the papers

Which may not be eaten?

Which may not be worn?

Which is only illusion?

Man always ignores

The wonders of nature

By forgetting natural attributes;

Praises his artificiality,

Believes science instead of nature,

Forgets the greatness of nature

Man is unable to make

Himself equal to nature

Or great as nature

Can't be equal with nature

He is part of nature.

How is he a part of nature?

Can he possess the whole of nature?

How will you explain?

The secret of relations

Between nature

How will your science explain?

Relations that are the life

Of every human being,

The existence of human beings,

The vitality of human beings

Those relations of nature

Create the whole creation

Vegetation, plants and flora,

All living beings over the earth

Which required no contribution?

Of human science!

Science is unable to reach

The depths of relations

Between elements of nature

Human existence will vanish

Before finding depths of relations

Man has not such a capability

To discover the depth

Of relations of natural elements

Using the strength of his science

Every moment those relations

Serve the whole creation,

Never deceitful when serving

Nature performs its relations

Without any hypocrisy,

But man deceives man

At every step, all the time,

Discriminates everywhere openly

Science is full of discrimination,

Playing a few hands

Human motive is deceitful

Not selfless like nature;

Human law of the land

Is full of discrimination

Full of deceitfulness

Works under the shadow

Of bombs and ammunition,

Has to eat bullets by the mouth

If anyone agitates against law

Nature needs no such bullets

To abide natural law

If man is not fair

Towards another human being,

How can he be fair to the earth?

Which nourishes all creation?

To which taking in possession,

Doing atrocity on humans,

How will he be the friend of the air?

A friend of water and sky,

Who is not friend of man?

How can he be of nature?

How can he be of living being?

How can he be of flora?

The man, who is not able

To understand his own existence

His own doings

How can he understand?

He has no sense or courage

To understand great nature

Now tell if you can

What is the relationship?

Of the Earth and the Sun

Explain this to the whole world.

You look very eager

To be the driver of nature

To be guard of the earth

Express yourself about the sun,

Tell its relation with earth.

You took the earth in possession

How will you please the Sun?

How will you visit the Sun?

To know its relation

With this beautiful Earth

How will you give your decision?

Or will give an ex parte decision

With your imperfect knowledge

How will you understand?

The deepness of relations

Earth is not rotating without reason.

Around its axis

Not revolving around the sun

Without any purpose

There should be some reason,

There should be some fundamental

In the relation of Earth and Sun

They must help each other

There should be some contribution

For each other to perform duties,

To construct their relationship

They must have some dependency

On each other, the Sun and Earth

To do their routine function,

To fulfill their relationship

Is science able to tell?

Who helps whom?

In rotating Earth around its axis

Or revolving around the sun

Or running the whole system

Of the whole of nature

Is man able to tell?

What is the relationship?

Between the water and the earth

They are together just for nothing?

Or do they have other deep ties?

Who helps whom?

Why water wanders in the sky,

Over casting sky with clouds,

Sometimes by becoming a stream,

Sometime becoming snow fall

On the mountains and plains

Sometime falling through rain,

Flowing through rivers

On and in the earth

To nourish the creation

By falling rain everywhere

Rivers fall in the sea

And again becomes steam

Again clouds and rain

Ocean on and under the earth,

How nature is related to each other,

Creating and nourishing creation.

Whether or not man understands

The depth of relationships

Who is their creator?

What is their power?

What type of engine is working?

For millions and billions of years,

Water and earth together

 Nourishing their creation,

Together maintaining velocity

Keeping them in time

By rotating and revolving.

Has the man or his science

Come to know the secrets?

Has the man any sense

To tell about relations of nature,

Of which he is eager

To catch the rein

Does he understand?

The nonstop routine of nature,

Unbreakable relations

Of sun and earth,

Relations of earth and sky,

Water and earth

Who fix the distance?

Between sun and earth!

How much distance is necessary!

How much warmth required!

How much water needed!

Does man know the requirement?

Of each other's cooperation?

Does he understand harmony?

Overall systems of nature

Has the man estimated?

How much share of sun,

How much share of moon,

How much share of earth,

Of water and air,

How much share of stars

Are in the open sky?

And what is the role of man

In the system of nature

If he understands nature,

He should explain to the world

And get the recognition.

If he is able to run

Any part of nature's system,

Is he able to run the earth?

Is he able to build such an engine?

An engine that is able do earth's work,

Able to get the support

Of whole nature

Through the ages

Earth remains the battlefield,

And also the treasure

Of its beautiful colors

Nature knows no difference,

May be a win or defeat

It remains the one who nourishes

The whole of creation

All living beings are its creation,

Maybe they eat each other,

But living beings are souls.

Some come, some go,

Some make the universe beautiful,

But some destroy

Other living beings

Over the bosom of the earth

This is the exhibitionist of power!

None leave stones unturned,

Everyone treats himself as brave

In this colorful nature

All put forth their share,

Some sing its song

Some enjoy its fruit

Some die in grief

Play of wins and defeats

Some show power

Some show greatness

Some die in ego

Some bow before power

Victory after victory

Man started taking dreams

In every age

To see their victory over nature,

To dominate nature

Does the pride of his power

Of his knowledge

Of his science

And of the development

Of his advanced science

In the era of every age

Sometimes the rain of bombs,

Sometimes the wrath of Astra Shastra

In every age,

Man remained eager

To become the owner of nature,

To take charge of air

To keep control of water

To take possession of sky

To take control of the warmth

Of the sun as a whole

To open the secrets of the moon

To declare himself king

Of the whole sky

To remove the name of nature

To write his name

In every age well built

The vanity of humans

Man supported the cause of man

To be great with materiality

Forgetting the sacred cause of nature

Man to be slave of man.

All the time man forgets

Nature's almighty,

The blessings of nature

In the end, every time

 Man's vanity is broken,

Obliterating the name

Of his knowledge from earth,

And of his science,

Including his progress

Nature realized man,

The strength of his power,

Recognition of its existence

Defeating the vanity of man

In all ages

Worthless greed never won,

Lustful man never won.

In the end, which atrocity did he enact?

To incinerate the earth

His knowledge is mortal,

A dead end

Nobody knows this.

How many times has he ruined

From the hands of nature

How many times vanished

Knowledge in the hands of nature

Always, nature made man

From intellectual to ignorant

He stands all the time

To see his own face,

Gathering little by little

Man again started to construct

The castles of knowledge,

And becomes addicted to the same

 New ambience,

To do massacres, Make weapons,

Bombs to atom bombs

Then hydrogen bombs

And chemical bombs,

Or makes missiles

And develop drones.

Maybe in some previous age

The names of these instruments

May be called by other names

May be Astra or Shastra?

May be bomb?

May be P*ashupatastra*?

Or may be *Maheshawarastra*,

Brahmshira or *Brahmaastra*?

May be *Bhargavastra*?

Vajra, *Mohini* or *Twashtar*?

May be developed something other?

Working like the drone of today

Otherwise nobody can become

Anything, just for nothing,

Son of water or sun of air

Man cannot fly with mind power.

Has to develop something to fly,

Which must be developed in that age?

Because man kept mouth

Upwards all the time

Satellite, computer, TV

Are not the adventures

Only of the present age

Sanjay had also been narrating

Live telecast of the Mahabharta War

To the blind king Dhritrastra,

Sitting with him in his palace

That was the adventure

And development of that age,

Which have left no sign behind?

Which finished completely?

Today's knowledge is totally new.

If science of the Mahabharta time

Could not be found in any form,

How we can presume today's

Present missiles, atom bombs

Satellites and drones

Will remain forever on earth

To perpetrate wrath

Knowledge is mortal.

Nature is always the nourisher,

Will remain nourishing the creation

As it did in the past,

To every living being,

Vegetables, plants and flora

Without any disparity

Man looks determined

To achieve victory over nature;

Wastes no time

To show nature as meek,

To give some shape

To his evil intentions

Working day and night,

Making the clones

Doing the experiments

To produce artificial man,

To produce artificial living beings,

Discovering such cells

And putting a stamp on it

Man made by science,

Man produced without nature

From the knowledge of man

Now he does not need nature

To give birth to man,

And is likely to declare soon,

To be revolutionary against nature

To raise flags against nature;

Started preaching openly

Of his unlimited vanity

Which is the end of his knowledge?

Man is an image of five elements,

He will make alternate living beings.

Actually man is an effigy of five elements

In which god is living!

And becomes a self-devotee

Man self is everything.

The form and appearance of nature,

Beautiful harmony of nature

But ensnaring his thinking

Producing some toys

Thinking himself powerful

Becoming very great

Showing himself equal to nature

Talking of himself as of nature

Wants to be driver of nature

Forgetting his pusillanimity

Unable to produce his own blood

Unable to make his own skin

Could not make his bones

Thinking to make a complete body

By the way, what will give an answer!

To men born naturally,

Why make bloodshed!

Why kill half the world,

Preventing natural sources

From helping the bane of hunger,

Dying with diseases

Without proper cure

Somewhere the elixir of life

Somewhere the rivers of milk

Somewhere the rivers of blood

Discrimination at every step

Everywhere a trader-ship

What more will he do!

By making more clones

Is he becoming able to run?

The playing with big nature,

Is he able to create?

Such a big creation

Or is proving himself

A great fool?

How beautiful life is

On the beautiful earth

Splendid flora and vegetation

On the winsome earth

Nourishment by nature

Of its own creation,

Passing through ages to ages

In the same rhythm,

Every moment, day and season

Nature never stops serving.

No difference to nature

If you are waking or sleeping

Mother Nature is great.

Everything has its own existence,

Eternity of nature is in everything

No difference if a man or dog

The secrets nobody knows:

Who runs this universe!

How everything works,

What is the strength of it!

Hustle and bustle over earth

Enjoying every living being,

None is able to understand.

Mother unworthy or son,

Always man's expectations are

To be driver of Mother Nature

Nature living in living beings,

Vegetation, plants and flora

Self explainable everywhere

Maybe time is difficult

But nature's work is guileless,

Bearing the curse of interference

Of man and science

Hill of ants is happy

Or crowd of men

Bevy of sparrows

Or lines of cranes are happy

Who enjoys more life?

Who will tell?

Air conditioning of science

Man becoming habitual,

Never tired of praising himself,

Claiming the extreme of science,

Convincing the world with "miracles",

Loitering with pump and show

Showing the nature trivial,

Praising his achievements

All available in the market

With his paper money,

Telling the importance of papers

Discriminating with paper money

By the man with the man

Openly looting the people,

The meaning of his power

Man telling man,

This ill-will wants to enforce

The rule of nature

Sees its greatness in it

But could not see

The blessings of great nature,

Inexhaustible air everywhere

Over the earth in every corner

Air never stops working.

It is the breath of man,

It is the life of living beings,

Air is the element of body

Saves us from hot weather

Sometimes a puff of air

Gives relief from hot weather

Gives enjoyment everywhere

Living in every living being

Keeps the body hale and hearty

Air is the element of body

Keeping life moving,

Breathing of every living being

It is the life to all,

Of vegetation and flora

Pure air for living beings;

Dirty air is the feed of flora

Surrounding everywhere

Nourishing whole world

Never discriminate when serving

With any living being

No partiality, no favoritism.

The part of its servings

Serves the whole of creation guilelessly

Science is not able to build

Such a big air conditioner

Man has not and shall never have

Such a big air conditioner,

To serve the whole of creation,

Free of cost, guilelessly.

With some motors and machines

He is able to make ice,

Or able to serve cold drinks

Means he has not become so great

By making ice from water

He forgets all the time

Nature produces enough water

Without any hurdle or machine,

Serving the whole of creation

For millions and billions years

With oceans and seas

Sunshine makes steam and clouds

Those wander in sky with air

To nourish the creation with rain

To run through streams and rivers

On and under the earth,

In oceans on and under the earth

Somewhere snow falls.

Somewhere there are glaciers of snow.

Somewhere water freezes.

Somewhere water is melting into water.

Somewhere water is boiling.

Somewhere a person is drinking water.

Somewhere a person is sweating profusely.

Everywhere, the glory of water,

Wandering everywhere

To nourish living beings,

Vegetation and flora

It visits every corner

Not having any selfness

And it charges nothing for services

Giving tireless services

Without any discrimination

Science has nothing equal to it

Nor is it able to serve like that,

Except for otiose vanity

Which is eager to rule?

Over all of nature

Science, which is ignorant,

How can it work on a large scale?

Who controls everything?

How does creation come and go?

From where does it come?

Where does it go?

Water has different forms.

How and who makes seasons?

Whose hand has control of it?

Who is the creator of all this?

Why and how is water nourishing?

How is the Earth rotating so punctually?

Around its wonderful axis

How is it revolving tirelessly?

Around the Sun in the sky

Who is controller of all this?

Man, who is feeling proud

Of his research of few machines

But research of computers,

Research of satellites and cameras

Which are not worthy for nature.

Nor are these discoveries able

To run any of element of nature

That is equal to the work of nature.

Nature functions methodically,

Recording the account of

Every minute and second

Of the velocity of Earth

Nothing in science is able to run

Without installing engines

How man come to the conclusion,

Earth is wandering in the sky

Without any meaningful cause

For rotating around its axis

And revolving around a sun,

How does it maintain punctuality?

How is it rotating and revolving?

How is it moving?

How does it remain in time?

For millions and billions of years

In the same routine

After every day comes night,

After every summer comes winter.

It is never entrapped

In black holes, discovered by man

Never came late or advanced

From a fixed time

By the way, how is it possible?

Without any type of engine

Which man has not yet discovered

Or is it impossible to discover

Or could not find due to fear

Or reverse action

Or know his limits.

How it is possible he is ignorant

And doesn't expect the truth?

There is an engine in the earth

Fully controlled by earth,

Working together with nature

There are burning furnaces

All the time in the earth

To run the engine of earth

Making volcano, lava and tides,

Boiling oceans in the earth

To provide regular steam

To the engine of earth

There are treasures of metals,

Treasures of oils and gases

All are feeding the earth

To give energy to the earth

To the system's punctiliousness

All elements of nature,

Earth and water,

Fire, air and sky

Work together to help each other;

Nothing is likely to happen

By looking upwards towards the sky

Nothing can be ignored

By shutting our eyes from truth

Reality will always prevail.

It is the play of nature,

Open before the universe

Nothing is hidden from anyone

No one can run from the naked truth.

Impossible to search mother's stomach,

Unable to understand mother's stomach

How Earth is rotating and revolving,

How its system is working,

Nobody able to understand

No such other system

Could be prepared by science

The truth, where burning furnaces

Where regularly boiling water,

Where burning gases, oil, metals

One day will come ultimately.

When that will become wasteful,

And will be an unnecessary burden on the earth.

It will become difficult to maintain balance

When the climate will be rebellious;

Excessive hot and cold

When there will be great turmoil

Unbearable for living beings,

For vegetation and flora,

When there will be lamentation

When there will be topsy-turvy

Again and again on earth

Earth will also be perturbed.

Earth could not heal living beings.

Knowledge will be useless.

Science will be helpless.

Nature and earth have to act.

The wasteful part of earth has to be separated,

Which is a very difficult process?

When nature will be in the process

To take that difficult step,

Knowledge will vanish in it,

No sign of science will remain.

The inner part of the earth will come out

And the outside of earth will go in

To get new treasures of energy

This will be the end of knowledge

The end of human science

All this will vanish in this process.

There will be no estimate.

What has vanished?

And what remains in existence

During that period of topsy-turvy

Man has to start again

The new story of knowledge,

A new start from the beginning,

Maybe he will see in the future

Two moons in the open sky.

May some new rotation

Of moons and earth

In the open firmament

Make a social animal again?

It has to start from a Stone Age

Or under compelling circumstances

It has to stand to keep alive

In the category of animals